D1085286

ADVANCES IN CYCLIC NUCLEOTIDE RESEARCH
VOLUME 2:
NEW ASSAY METHODS FOR
CYCLIC NUCLEOTIDES

Advances in Cyclic Nucleotide Research

Series Editors:

Paul Greengard, New Haven, Conn., U.S.A.
G. Alan Robison, Houston, Texas, U.S.A.

International Advisory Board

Bruce Breckenridge, New Brunswick, N.J., U.S.A.
R. W. Butcher, Worcester, Mass., U.S.A.
E. Costa, Washington, D.C., U.S.A.
Nelson Goldberg, Minneapolis, Minn., U.S.A.
Otto Hechter, Chicago, Ill., U.S.A.
David M. Kipnis, St. Louis, Mo., U.S.A.
Edwin G. Krebs, Davis, Cal., U.S.A.
Thomas A. Langan, Denver, Col., U.S.A.
Joseph Larner, Charlottesville, Va., U.S.A.
Grant W. Liddle, Nashville, Tenn., U.S.A.
Yasutomi Nishizuka, Kobe, Japan
Ira H. Pastan, Bethesda, Md., U.S.A.
Th. Posternak, Geneva, Switzerland
Theodore W. Rall, Cleveland, Ohio, U.S.A.
Martin Rodbell, Bethesda, Md., U.S.A.
Charles G. Smith, New Brunswick, N.J., U.S.A.

ADVANCES IN CYCLIC NUCLEOTIDE RESEARCH

Volume 2:

New Assay Methods for

Cyclic Nucleotides

SERIES EDITORS:

Paul Greengard, Ph.D.
Professor of Pharmacology
Yale University School of Medicine

G. Alan Robison, Ph.D.
Professor of Pharmacology, and
Director of the Program in Pharmacology
University of Texas Medical School at Houston

VOLUME EDITORS:

Paul Greengard, Ph.D.
New Haven, Connecticut

G. Alan Robison, Ph.D.
Houston, Texas

Rodolfo Paoletti, M.D.
Milan, Italy

Raven Press, Publishers · New York

© 1972 by Raven Press Books, Ltd. All rights reserved. This book is protected by copyright. No part of it may be duplicated or reproduced in any manner without written permission from the publisher.

Made in the United States of America.

International Standard Book Number 0-911216-21-9
Library of Congress Catalog Card Number 71-181305

Preface

This second volume of *Advances in Cyclic Nucleotide Research*, like the first, is a result of an international conference on cyclic AMP held in Milan during the summer of 1971. On the afternoon of the first day of that conference a panel discussion on methodology was held. Following an introductory lecture by Dr. A. G. Gilman, a variety of other advances in methodology were presented and discussed. Many of these methods had been published previously, but were widely scattered throughout the literature. Furthermore, almost all of the participants indicated that improvements had been made since their initial publication. Before the afternoon was over it became clear to us that an extremely useful volume could be based on this discussion. Each of the panelists was therefore commissioned to prepare a complete description of the method which he had presented, and to do so in such a way that the method could be readily followed in other laboratories. All the panelists generously agreed to undertake this task, and all of them executed it with their customary thoroughness and clarity. The result is a handbook of methods in cyclic nucleotide research that we believe will remain valuable for many years to come. It should be useful to all workers in the field, but especially to those who have had little previous experience in it.

Readers concerned about the rapid rate of progress in this area can be assured that when better methods are devised they will be promptly and thoroughly covered in future volumes of this series.

Paul Greengard
Rodolfo Paoletti
G. Alan Robison

Contents

Advances in Cyclic Nucleotide Research, Vol. 2
Raven Press, New York © 1972

Protein Binding Assays for Cyclic Nucleotides

Alfred G. Gilman

*Department of Pharmacology, School of Medicine, University of Virginia,
Charlottesville, Virginia 22903*

I. INTRODUCTION

The diversity of assays available for adenosine 3′,5′-cyclic monophosphate (cAMP) and guanosine 3′,5′-cyclic monophosphate (cGMP) is both impressive and highly desirable. Analytical difficulties are imposed by extremely low tissue levels of the cyclic nucleotides and considerably higher levels of potentially interfering compounds. The existence of several good techniques for estimation of quantities of cyclic nucleotides in biological material ensures the fact that different laboratories will be subject to different sets of limitations and possible errors. The ultimate result should thus be beneficial.

Most recently, assays for cyclic nucleotides have been developed which are based on competition for protein binding of the compounds (Gilman, 1970; Walton and Garren, 1970; Brown, Albano, Ekins, Sqherzi, and Tampion, 1971; Murad, Manganiello, and Vaughan, 1971; Murad and Gilman, 1971). Such methods have become feasible because of the discovery of proteins that bind cyclic nucleotides with high affinity and because of the development of simple techniques for separation of bound and free nucleotide. The method for cAMP determination to be described in detail in this chapter utilizes a binding protein that is presumably a cAMP-dependent protein kinase (Walsh, Perkins, and Krebs, 1968). Impetus for the development of the assay was provided by the discovery of Walsh, Ashby, Gonzalez, Calkins, Fischer, and Krebs (1971) that a heat-stable protein inhibitor of the protein kinase (Posner, Hammermeister, Bratvold, and Krebs, 1964; Appleman, Birnbaumer, and Torres, 1966) *increased* the affinity of cAMP for the enzyme. A protein kinase from muscle was chosen because of its favorable binding constant for cAMP (Walsh et al., 1971) and because of the availability of large amounts of tissue.

It was determined that both cAMP-dependent protein kinase and cAMP-binding activities from muscle extracts could be quantitatively adsorbed on cellulose ester (Millipore) filters. This filtration method is an extremely simple means for separation of bound from free ligand and thus forms the basis for estimation of the extent of competition for binding sites between ^{3}H-cAMP of high specific activity and unlabeled cAMP to be assayed.

A method for the determination of cGMP has been developed by Murad et al. (1971) that is identical in principal and nearly identical in operation to the cAMP binding assay. For cGMP, however, a cGMP-dependent protein kinase from lobster muscle is utilized (Kuo and Greengard, 1970). In addition, we have recently determined that the cAMP and cGMP assays can be combined in a single test tube and performed simultaneously to assay the concentrations of both nucleotides in urine and perhaps in certain tissues (Murad and Gilman, 1971).

II. CYCLIC AMP ASSAY

A. Binding Protein

While a more extensive purification scheme was utilized to characterize the protein and to establish its co-chromatography with the cAMP-dependent protein kinase, a simplified method is more than sufficient for routine use. Fresh bovine muscle[1] was prepared as described by Miyamoto, Kuo, and Greengard (1969) through the ammonium sulfate precipitation step. Briefly, this involved homogenization, centrifugation, pH 4.8 precipitation, and ammonium sulfate precipitation. This preparation contains two major peaks of kinase and cAMP binding activities and is fractionated on DEAE-cellulose. For example, a dissolved and dialyzed ammonium sulfate precipitate derived from 250 g of muscle was applied to a column of Whatman DE11 (1 meq/g, 32 × 2.6 cm) previously equilibrated with 5 mM potassium phosphate, pH 7, and the column was washed with this buffer. Kinase and binding activities were then eluted in one of two ways. Elution with a linear gradient of potassium phosphate, pH 7, from 5 to 400 mM is shown in Fig. 1. Alternatively, the first peak of activity can be eluted with 100 mM potassium phosphate, pH 7, and the second peak collected with the same buffer at 300 mM concentration. The second peak is dialyzed against 5 mM potassium phosphate, pH 7, and is used for the assay. Binding activity has been completely stable for 18 months at − 20°C. The "batch" elution method is slightly easier; however, gradient elution is undoubtedly more foolproof.

---

[1]Different investigators have had somewhat variable experience with muscle obtained from commercial sources. Freshly slaughtered material has occasionally had poor activity. By contrast, we and others have obtained preparations with as good or significantly higher specific activities than those reported by using muscle (shoulder and rump steaks) of indeterminate age from butcher shops.

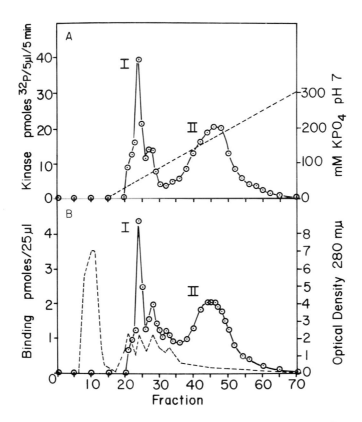

FIG. 1. DEAE-cellulose chromatography of protein kinase and cAMP binding activities. Gradient elution was performed with potassium phosphate, pH 7, from 5 to 400 mM. Fractions collected were 20 ml. Binding activity was assayed at pH 4 with 40 nM cAMP. Kinase activity was assayed as described by Miyamoto et al. (1969) with histone as substrate. A: o, protein kinase activity; --, potassium phosphate gradient. B: o, cAMP binding activity; --, optical density at 280 nm.

While the binding protein can be utilized in the assay at almost any stage of purity, some increase in sensitivity is obtained by separation of the two peaks. Thus the pH optima for binding of cAMP are somewhat different for the two activities, and the binding constant can be optimized only if they are separated. The entire purification procedure requires two days. As the yield of binding protein from 500 to 1,000 g of muscle is sufficient for more than 10^5 assay tubes, this expenditure of time seems thoroughly insignificant.

The binding protein preparation utilized here bound 0.3 pmoles cAMP/μg protein under cAMP assay conditions. Over 200 μg of this protein could be quantitatively adsorbed by a single Millipore filter.

B. Protein Kinase Inhibitor

The inhibitor preparation was modeled after that of Appleman et al. (1966). Bovine muscle was homogenized in 10 mM Tris chloride, pH 7.5, and was boiled for 10 min. After removal of particulate material by filtration, activity was precipitated with 1/9 volume of 50% trichloroacetic acid. The precipitate was collected at 15,000 $\times$ g, dissolved in water, and the pH was adjusted to 7 with 1 N NaOH. This fraction was dialyzed against distilled water at room temperature, and the copious gelatenous precipitate which formed was discarded. The inhibitor preparation may be assayed in the cAMP-dependent protein kinase reaction (recommended) (Walsh et al., 1971) or by its effect of enhancing the affinity of cAMP for the binding protein.

C. Cyclic AMP Binding and Cyclic AMP Assay

An individual decision was made to conduct the binding reaction for cAMP assay under saturating conditions of the nucleotide. Although some sensitivity is lost, interference is minimized and analysis of data is facilitated. The reaction volume is kept small to attain relatively high concentrations of ^{3}H-cAMP. While the total volume and total amount of ^{3}H-cAMP used may readily be individualized to meet specific needs, changes in cAMP *concentration* (to subsaturating levels) may cause potential interference to become significant.

For the purpose of cAMP assay, the binding reaction is conducted in a volume of 50 μl of 50 mM sodium acetate/acetic acid, pH 4.0. The only other reaction components are ^{3}H-cAMP (Schwartz Bioresearch, 16.3 C/mMole)[2], unknowns or standard cAMP solutions, sufficient binding protein to bind less than 30% of the nucleotide[3], and, where indicated, a maximally effective concentration of the protein kinase inhibitor preparation. Reactions are initiated by addition of binding protein after chilling of tubes and are allowed to proceed for greater than 60 min at 0°C. At equilibrium, the mixtures are diluted approximately 1 ml with cold 20 mM potassium phosphate, pH 6, and are passed through a 24 mm cellulose ester Millipore filter (0.45 μ) previously rinsed with the same buffer. The filter is immediately washed with 5 to 10 ml of this buffer and placed in a counting vial with 1 ml of methyl cellosolve, in which the filter readily dissolves. Finally, a scintillation cocktail of toluene: methyl cellosolve (3:1) plus fluors is utilized. Bound counts are independent of filtration speed and the volume of rinse from 3 to 20 ml. In the absence of binding protein, $\sim$ 20 cpm are counted on the filter.

---

[2]More recently a generally labeled ^{3}H-cAMP (24.1 C/mmole, New England Nuclear) has proven to be highly satisfactory.

[3]This limit for percentage of ^{3}H-cAMP bound was set arbitrarily. It was desired to maintain saturation with a generous excess of free cAMP.

The concentration of ^{3}H-cAMP in the assay is somewhat arbitrary. For the assay of cAMP, ^{3}H-cAMP is utilized at 10 to 20 nM (0.5 to 1.0 pmoles/50μl) in the presence of the inhibitor or 40 nM in its absence. As these are saturating concentrations of ^{3}H-cAMP, the effect of added unknown or standard cAMP solutions can be evaluated from a nearly theoretical decrease in the total bound ^{3}H-cAMP, which is entirely linear when plotted logarithmically.

D. Tissue Extracts

Frozen tissue samples are homogenized in 1 ml of cold 5% trichloroacetic acid. TCA supernatants are extracted 5 times with 3 volumes of water-saturated ether after the addition of 0.1 ml of 1 N HCl. Residual ether is removed at 80°C for 1 to 2 min, and the aqueous extracts are lyophylized and redissolved in 50 mM sodium acetate, pH 4.5. Single aliquots of such extracts are usually assayed at two or three dilutions over a fourfold concentration range.

E. Assay Characteristics

1. pH Optimum and Binding Constant. The effect of pH on binding is critical and has been investigated in some detail (Gilman, 1970). Briefly, the highest affinity seen was at pH 4.0 in acetate buffer. Binding at very low cAMP concentrations is significantly reduced above this pH, and binding is drastically altered below pH 3.5. Adequate control of the pH of samples to be assayed is thus stressed.

Data from which the binding constant may be evaluated are presented in Fig. 2. While the binding constant is 10 to 20 nM at pH 6 (phosphate buffer), a marked increase in affinity (K_a = 2 to 3 nM) is seen at pH 4. In the presence of the inhibitor preparation, the binding constant approaches 1 nM.

2. Kinetics of Binding. At 0°C, binding equilibrium is established within 60 min, and in the presence of the inhibitor preparation the equilibrium plateau is completely stable for at least 4 hr. Without the inhibitor preparation, a slow decline (approximately 5% per hr) in bound counts is apparent.

The reverse reaction (studied either by dilution or by addition of excess unlabeled cAMP) is very slow and is first order. In the absence of inhibitor the $t_{1/2}$ for complex dissociation was approximately 7 hr at 0°C, and the reaction was somewhat slower in its presence. In addition, without inhibitor approximately 10% of the bound counts exchanged rapidly after dilution. For this reason, samples were filtered 3 to 4 min after dilution when inhibitor was not used. This waiting time is not necessary when the inhibitor is present.

Several effects of the inhibitor preparation have thus been noted. In general, there appears to be increased stability and a somewhat greater number of total binding sites, in addition to the increased affinity. It now appears that

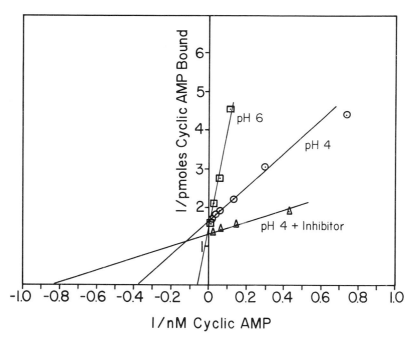

FIG. 2. Estimation of cAMP binding constant. Binding was determined at pH 6 (50 mM potassium phosphate) (□) and at pH 4 in the absence (○) or in the presence of 45 μg inhibitor fraction protein/200 μl (Δ). Binding protein concentration was 2 μg/200μl.

at least some of the phenomena attributable to the inhibitor preparation are nonspecific (Murad and Gilman, 1971). Thus, when low amounts of binding protein are present, either inhibitor preparation or albumin appears to enhance stability and the total number of binding sites available. While the effect of the inhibitor on affinity is presumably specific, detailed study on the effects of other proteins has not been undertaken.

3. Competitors. Table 1 demonstrates the effect of nucleotides and related compounds on cAMP binding. As might be expected, other 3′,5′-cyclic nucleotides are more effective. However, mammalian tissue levels of cyclic GMP are not sufficiently high to interfere. Of pragmatic significance is the fact that ATP had only a 50% inhibitory effect at 1 mM and virtually no effect at 0.1 mM. Effects of this nucleotide in tissue extracts thus should not be seen if extracts are assayed at ultimate dilutions of 10- to 50-fold (depending on the tissue). Similar effects of these compounds are observed when examined with 20 nM cAMP in the presence of the inhibitor preparation.

TABLE 1. *Effect of nucleotides and related compounds on cAMP binding*

Compound	Concentration (μM) at 50% inhibition[a]	Compound	% inhibition at 1 mM[a]
cIMP	0.3	UTP	30
cGMP	5.0	CTP	28
cUMP	10	5'-AMP	21
cCMP	30	ADP	18
GTP	700	adenosine	0
ATP	1,000	theophylline	0

[a]cAMP concentration = 40 nM; binding protein = 2 μg/200 μl.

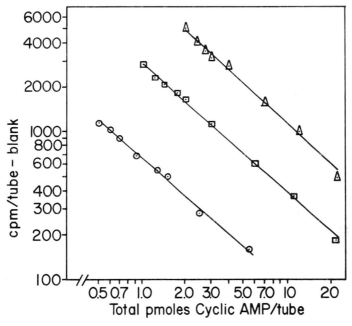

FIG. 3. Standard curves for cAMP assay. All reactions were carried out at pH 4 and 0° in a volume of 50 μl. ^{3}H-cAMP added/tube was ⊙, 0.5 pmole; ▫, 1.0 pmole; Δ, 2.0 pmole. 14 μg protein of the inhibitor fraction was present at the two lower cAMP concentrations, and binding protein was added at 0.5, 1.0, and 2.0 μg for the three conditions, respectively. Known quantities of cAMP were added to achieve the total (labeled plus unlabeled) indicated content of cAMP/tube.

4. Standard Curves. Standard curves obtained with *saturating* concentrations of ^{3}H-cAMP are presented in Fig. 3. Total pmoles of cAMP plotted on the abscissa represents the sum of ^{3}H-cAMP and unlabeled standard added per tube. Thus a variety of curves can be generated, depending on the sensi-

TABLE 2. *cAMP levels in mouse liver and brain*

Tissue	Sample and conditions	cAMP (pmole/mg wet wt.)
liver[a]	2 mg	1.0
	10 mg	0.9
	10 mg + 1 pmole/mg cAMP	2.0
	10 mg + phosphodiesterase	0.0
brain[b]	2 mg	2.1
	10 mg	2.1
	10 mg + 1 pmole/mg cAMP	3.2
brain[a] (15 sec decapitation)	2 mg	6.8
	10 mg	6.3
	10 mg + 1 pmole/mg cAMP	7.3
	10 mg + phosphodiesterase	0.0
brain[a] (120 sec decapitation)	5 mg	18

[a]Tissues were removed from decapitated mice and frozen in liquid nitrogen.

[b]Mice were sacrificed by immersion in liquid nitrogen and frozen brain tissue was dissected.

tivity required. They are straight lines throughout on a logarithmic plot as a result of the condition of saturation, and they are very close to the theoretical slope predicted for this condition. With the most sensitive curve shown, a 20% reduction of total cpm bound is obtained with the addition of 0.10 pmole unlabeled cAMP. Further reduction of volume and/or the amount of ^{3}H-cAMP added will readily result in an assay sensitive to < 0.05 pmole/tube and one in which the specific activity of the ^{3}H-cAMP available is becoming the limiting factor.

5. *Assay Specificity.* Representative assay data are shown in Table 2, and tissue levels obtained are in reasonable agreement with those in the literature. More importantly, the independence of the values from the amount of tissue analyzed (with final tissue dilution in the assay of 25- to 250-fold) and the results of cyclic 3',5'-nucleotide phosphodiesterase treatment indicate the specificity of the procedure. More data on specificity will be shown below. Finally, known amounts of cAMP added at the time of homogenization were quantitatively recovered.

III. CYCLIC GMP

The assay for cGMP has been developed by Murad et al. (1971).

A. Binding Protein

The cGMP binding protein is prepared from lobster tail muscle through the dialyzed ammonium sulfate precipitation step described by Kuo and Greengard (1970) for cGMP-dependent protein kinase. While this preparation is rather crude, it is sufficient for assay purposes. Difficulty was encountered in attempts to purify the protein further. The binding activity is stable for months at $-70°C$, but activity is lost with repeated freezing and thawing.

B. Cyclic GMP Binding and Cyclic GMP Assay

This assay has also been described with saturating concentrations of cGMP. For cGMP assay, the binding reaction is usually conducted in a total volume of 100 μl of 50 mM sodium acetate/acetic acid, pH 4.0. Other reaction components are ^{3}H-cGMP (6–10 pmoles), unknowns or standard cGMP solutions, and cGMP binding protein (usually 260 μg of the preparation described). This amount of protein binds only a small fraction of the ^{3}H-cGMP present. The binding protein is added to initiate the reaction after the tubes have been chilled. Tubes are incubated at 0°C for 75 min, and reaction mixtures are then diluted, filtered, and counted as described for the cAMP assay.

C. Tissue Extracts

Tissue extracts are assayed either after preparation, as described above for cAMP assay, or after purification on small columns of Dowex-1-formate. This column procedure separates cAMP from cGMP with nearly quantitative recoveries of both nucleotides. In addition, ATP is retained on the column and 5'-AMP, which interferes in the cGMP assay, is separated from this cyclic nucleotide. Urine samples were assayed directly or after purification on such columns.

D. Assay Characteristics

1. pH Optimum and Binding Constant. As in the cAMP binding assay, the highest affinity seen for cGMP binding was observed at pH 4.0 in sodium acetate buffer. There appeared to be some variability, however, in the binding constant observed with different protein preparations. Values of 2 to 10 nM were observed. Data obtained with a preparation showing an apparent binding constant of 5 nM are shown in Fig. 4.

2. Kinetics. Binding equilibrium in this system is attained within 75 min at 0°C, and maximal binding is maintained for at least 45 min thereafter. The rate and nature of the reverse reaction have not been examined in detail.

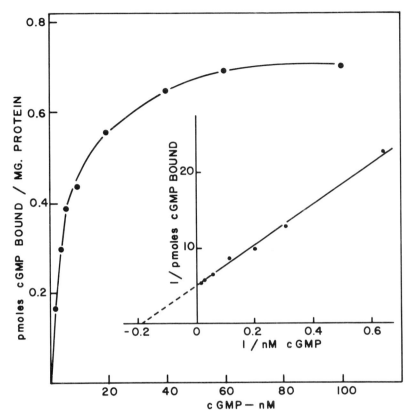

FIG. 4. Estimation of cGMP binding constant. Samples containing various concentrations of ³H-cGMP were incubated at pH 4 and 0° with 260 μg of binding protein/100 μl. (From Murad et al., 1971.)

3. Competitors. Cyclic AMP interferes with cGMP binding 15 to 20% only when present in a 10-fold excess. Equimolar concentrations of cAMP have little effect. Of numerous other naturally occurring nucleotides and nucleosides tested, only 5'-AMP inhibits significantly at 0.1 mM (15%) and 1.0 mM (30%) concentrations.

4. Standard Curves. A variety of standard curves for cGMP are shown in Fig. 5. Again, when plotted logarithmically, lines are straight, parallel, and close to theoretical. Such curves allow the estimation of 0.5 to 1.0 pmole of cGMP. Considerably more counts could be bound with ³H-cGMP or binding protein of higher specific activity.

5. Assay Specificity. Specificity of this assay and of the cAMP assay were ascertained by determination of levels of cyclic nucleotides in tissue extracts

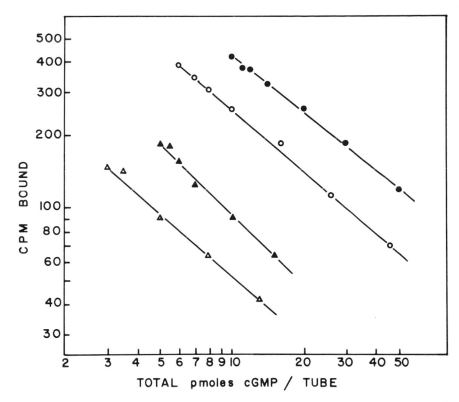

FIG. 5. Standard curves for cGMP assay. Tubes containing either 10 (●) or 6 (○) pmolc of
³H-cGMP in 100 μl or 5 (▲) or 3 (△) pmole in 50 μl were incubated as described. Known amounts
of unlabeled cGMP were added to some tubes as indicated. A 260-μg amount of binding protein
was used in 100 μl incubations and 130 μg of protein in 50 μl reactions. Total cGMP/tube
represents labeled plus unlabeled nucleotide. (From Murad et al., 1971.)

before and after purification on the Dowex-1-formate column described
(Table 3). Pre- and postpurification values for cAMP were in reasonably good
agreement for all tissues examined and for urine. Percentage differences are
slightly greater for the liver sample analyzed, perhaps due to the fact that
these samples, for somewhat mysterious reasons, contained cAMP levels
approximately fivefold lower than those expected and usually found.[4] When
unusual cAMP levels of < 2 pmoles/mg protein are encountered, some sample
purification would seem prudent, at least initially.

---

[4]The differences observed with the liver sample after purification are, however, in a direction
opposite to that expected if purification removes compounds that interfere with cAMP binding.
The larger percentage difference may thus represent greater analytical variability in a sample
with very low cAMP content.

TABLE 3. *Assay of cGMP and cAMP before and after purification of samples on an ion-exchange column*

Tissue	Sample	Assay (pmole/mg protein)			
		Cyclic GMP		Cyclic AMP	
		Before purification	After purification	Before purification	After purification
lung	1	3.2	2.4	5.8	6.2
	2	3.7	3.1	6.2	7.7
cerebellum	1	5.7	4.6	46.7	43.8
	2	2.9	2.6	34.6	28.4
liver	1	2.4	0.23	2.0	2.5
	2	3.3	0.06	0.8	1.2
		μmole excreted/12 hr			
human urine[a]	1	0.29	0.35	3.5	3.4
	2	0.72	0.73	3.5	3.1
	3	0.33	0.35	—	2.0

[a]From a study done in collaboration with Dr. C. Y. Pak

Urine samples and deproteinized rat tissue were prepared as described in the text and assayed for cyclic GMP and cyclic AMP content before and after purification on Dowex-AG 1 × 8 formate columns. Urine was collected in three consecutive 12-hr periods from a patient with hyperparathyroidism. At the beginning of the second collection period, the patient was given an intravenous infusion of calcium gluconate (15 mg Ca^{++}/kg) for 4 hr. (From Murad et al., 1971.)

The content of cGMP of unpurified extracts of lung, cerebellum, and urine agreed quite well with those found in purified samples. Samples from these sources, which contain rather high concentrations of cGMP, can thus be assayed without purification. However, tissues such as liver, containing extremely low quantities of cGMP, show a clear need for purification prior to assay.

IV. SIMULTANEOUS CYCLIC NUCLEOTIDE ASSAY

Reaction conditions and protein-nucleotide complex isolation procedures for the two assays described previously are essentially identical. Since each cyclic nucleotide produced little or no interference in the assay of the other, we attempted to develop a method for the simultaneous determination of both cAMP and cGMP in the same assay mixture. The double-isotope method employed and its application to cyclic nucleotide assay in urine have been described (Murad and Gilman, 1971).

The combined assay is typically conducted in 100 μl of 50 mM sodium acetate, pH 4.0. ^{3}H-cGMP (10 pmoles) and ^{32}P-cAMP (5 to 10 pmoles; ICN) are utilized. Somewhat increased amounts of ^{32}P-cAMP can be used as radioactive decay occurs. Tubes are chilled, and reactions are initiated by addition of a mixture of the two binding proteins. After incubation and filtration as described above, filters are counted for ^{3}H and ^{32}P content.

In the simultaneous (or individual) assay, cGMP does not interfere with cAMP binding until very high molar ratios are achieved (Fig. 6). Cyclic AMP produces a slight inhibition of cGMP binding at equimolar concentrations, but further inhibition is not observed up to a cAMP-cGMP ratio of 4:1. Unlabeled cAMP produces a nearly theoretical decrease in ^{32}P-cAMP bound, and unlabeled cGMP produces similar dilution of ^{3}H-cGMP bound in the combined assay (Fig. 6), just as shown previously when performed individually.

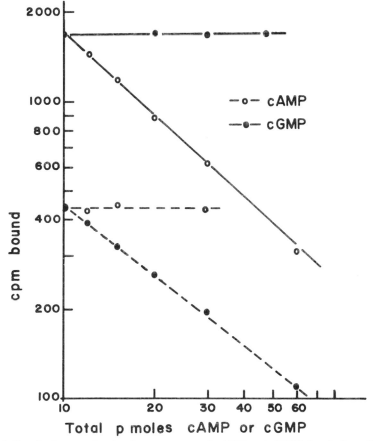

FIG. 6. Standard curves for simultaneous assay of cAMP and cGMP. Incubations (100 μl) were performed with 10 pmole ^{3}H-cGMP, 10 pmole ^{32}P-cAMP, 260 μg cGMP binding protein, and 1.3 μg cAMP binding protein. Solid lines represent amount of ^{32}P-cAMP bound and broken lines represent amount of ^{3}H-cAMP bound. To some tubes (—○ or --●--), standard solutions containing either 2, 5, 10, 20, or 50 pmole each of cAMP and cGMP were added. To other tubes standard solutions of only cAMP (2, 5, or 20 pmole, --○--) or cGMP (10, 20, or 40 pmole, ●) were added.

TABLE 4. *Cyclic nucleotide levels in human urine*

Nucleotide	Assay method	Urine nucleotide concentration (μM)			
		A	B	C	D
cAMP	individual	2.06	1.52	1.20	2.27
	combined	1.80	1.60	1.20	2.11
cGMP	individual	0.48	0.57	0.24	0.43
	combined	0.44	0.63	0.24	0.39

Unpurified urines (A-D in Table 4) were examined for cAMP and cGMP content using both individual and combined nucleotide assays. As summarized in Table 4, there was excellent agreement between the values obtained with the different methods.

V. DISCUSSION

The advantages of protein-binding assays for cyclic nucleotides are significant. Sensitivity, particularly for cAMP, is high. Specificity is sufficiently high that tissue extract purification seems unnecessary in most cases for cAMP and in some cases for cGMP. In the latter case, a simple and quantitative purification method can be coupled to the binding assay when necessary. The assays are exceedingly simple to perform, reaction mixtures contain few components, and reaction conditions are such that destruction of nucleotides is not a factor. Preparation of binding proteins presents no problems, and material for thousands of assay tubes is obtained in hours in both cases. All work required to obtain a functional assay can thus be performed within a few days.

Several further advantages arise when the assays are performed at saturation. Thus, standard "curves" are linear throughout when plotted logarithmically and are nearly theoretically predictable. Computer analysis is thus facilitated. Potential interference from competing materials is minimized and such potential variables as total volume become less critical.

It must be emphasized that when new biological systems are approached, appropriate controls for assay specificity must be performed. It is possible that some tissues will prove to require purification prior to cAMP assay and this will certainly be necessary in many cases for cGMP assay. When incubation media containing large amounts of salt must be concentrated prior to assay, desalting procedures are essential. Exogenous interfering compounds may also be added in experimental manipulations. In cases where purification (and consequent loss) of cAMP or cGMP is necessary, the situation may readily be approached, as in other assays, by the addition of ^{3}H-nucleotide (or other label) for the calculation of recoveries. Amounts of tracer may be utilized that are insignificant when carried into the assay reaction mixture.

Simultaneous assay of cAMP and cGMP in the same incubation offers obvious advantages. The method described permits detection of 1–2 pmoles of cAMP and cGMP in microliter samples of urine and can decrease technical time 50% when large numbers of urine samples are analyzed for cyclic nucleotides. While the applicability of the simultaneous binding assay for tissue extracts has not been investigated, similar dual radioimmunoassay will probably be more useful for simultaneous nucleotide assay of such samples because of the greater base specificity of the cyclic nucleotide antibodies (Steiner, Wehmann, and Parker, *this volume*). It would also seem likely that simultaneous displacement assays of a variety of combinations of compounds may be possible with different isotopic labels.

Acknowledgment. Work by the author was done while in the Laboratory of Biochemical Genetics, National Heart and Lung Institute, as a Pharmacology Research Associate, Pharmacology-Toxicology Program, National Institute of General Medical Sciences.

REFERENCES

Appleman, M. M., Birnbaumer, L., and Torres, H. N. (1966): Factors affecting the activity of muscle glycogen synthetase. III. The reaction with adenosine triphosphate, Mg^{++}, and cyclic 3',5'-adenosine monophosphate. *Archives of Biochemistry and Biophysics*, 116:39–43.

Brown, B. L., Albano, J. D. M., Ekins, R. P., Sqherzi, A. M., and Tampion, W. (1971): A simple and sensitive saturation assay method for the measurement of adenosine 3',5'-cyclic monophosphate. *Biochemical Journal*, 121:561–563.

Gilman, A. G. (1970): A protein binding assay for adenosine 3',5'-cyclic monophosphate. *Proceedings of the National Academy of Sciences*, 67:305–312.

Kuo, J. F., and Greengard, P. (1970): Cyclic nucleotide-dependent protein kinases. VI. Isolation and partial purification of a protein kinase activated by guanosine 3',5'-monophosphate. *Journal of Biological Chemistry*, 245:2493–2498.

Miyamoto, E., Kuo, J. F., and Greengard, P. (1969): Cyclic nucleotide-dependent protein kinases. III. Purification and properties of adenosine 3',5'-monophosphate-dependent protein kinase from bovine brain. *Journal of Biological Chemistry*, 244: 6395–6402.

Murad, F., and Gilman, A. G. (1971): Adenosine 3',5'-monophosphate and guanosine 3',5'-monophosphate: a simultaneous protein binding assay. *Biochimica et Biophysica Acta*, 252:397–400.

Murad, F., Manganiello, V., and Vaughan, M. (1971): A simple sensitive protein-binding assay for guanosine 3',5'-monophosphate. *Proceedings of the National Academy of Sciences*, 68:736–739.

Posner, J. B., Hammermeister, K. E., Bratvold, G. E., and Krebs, E. G. (1964): The assay of adenosine 3',5'-phosphate in skeletal muscle. *Biochemistry*, 3:1040–1044.

Steiner, A. L., Kipnis, D. M., Utiger, R., and Parker, C. (1969): Radioimmunoassay for the measurement of adenosine 3',5'-cyclic phosphate. *Proceedings of the National Academy of Sciences*, 64:367–373.

Steiner, A. L., Wehmann, R. E., Parker, C. W., and Kipnis, D. M. (1972): Radioimmunoassay for the measurement of cyclic nucleotides, *this volume*.

Walsh, D. A., Perkins, J. P., and Krebs, E. G. (1968): An adenosine 3′,5′-monophosphate-dependent protein kinase from rabbit skeletal muscle. *Journal of Biological Chemistry*, 243:3763–3765.

Walsh, D. A., Ashby, C. D., Gonzalez, C., Calkins, D., Fischer, E. H., and Krebs, E. G. (1971): Purification and characterization of a protein inhibitor of adenosine 3′,5′-monophosphate-dependent protein kinases. *Journal of Biological Chemistry*, 246:1977–1985.

Walton, G. M., and Garren, L. D. (1970): An assay for adenosine 3′,5′-cyclic monophosphate based on the association of the nucleotide with a partially purified binding protein. *Biochemistry*, 9:4223–4229.

Advances in Cyclic Nucleotide Research, Vol. 2
Raven Press, New York © 1972

Saturation Assay for Cyclic AMP Using Endogenous Binding Protein

B. L. Brown, R. P. Ekins, and J. D. M. Albano

The Institute of Nuclear Medicine, The Middlesex Hospital Medical School, London, England

I. INTRODUCTION

The recent introduction of saturation assay (i.e., radioimmunoassay; Steiner, Kipnis, Utiger, and Parker, 1969) and protein binding assay (Gilman, 1970; Walton and Garren, 1970; Brown, Ekins, and Tampion, 1970; Brown, Albano, Ekins, Sgherzi, and Tampion, 1971; Murad, Manganiello, and Vaughan, 1971) techniques for the measurement of cyclic nucleotides has opened the way to new areas of investigation in this field. The basic simplicity of these methods, coupled to the high order of sensitivity and specificity of which they are potentially capable, make them especially attractive. Nevertheless, there exist inevitable pitfalls associated with their use which must be avoided by newcomers to this type of analytical technique. Moreover, a misunderstanding of the fundamental, mathematical theory on which these techniques are based may prevent investigators from fully exploiting the sensitivities which the methods can, in principle, attain.

In this chapter we shall first describe briefly the steps in our own laboratory procedure for the measurement of cyclic AMP (Brown et al., 1971). It differs in detail from the method independently developed by Gilman (1970) in that a simple adrenal extract is used as the source of binding protein, and charcoal is employed to separate bound from free, labeled cyclic AMP.

The advantages, if any, that these differences hold over Gilman's technique are marginal and to some extent dependent upon the facilities available to the experimenter. The preparation of adrenal binding protein is technically simpler, and can be accomplished in 2 to 3 hr. However, since this step constitutes, in both methods, a minor part of the overall procedure, this advantage is relatively unimportant. Likewise, long experience in our laboratory with both membrane filtration and charcoal separation methods in the measurement

25

of other compounds has led us to rely almost exclusively on the charcoal technique, which in our experience is considerably quicker, cheaper, and equally precise. These essentially domestic advantages led us to adopt the charcoal separation method for measurement of cyclic nucleotides.

The differences between the naturally occurring protein saturation assay methods and those relying on antibodies are more fundamental in that the latter permit the use of iodine-labeled tyrosine cyclic nucleotide derivatives of higher specific activity in the assay. In certain circumstances this can lead to higher sensitivity. On the other hand, this approach requires fairly frequent preparation of tracer, and specific antisera have not been readily available. The circumstances in which it is advantageous to employ antisera (and iodine-labeled nucleotide derivative) rather than naturally occurring binding protein are not always obvious. It therefore seemed to us to be useful also to consider some of the theoretical factors which could influence the choice between assay methods relying on antibodies or cellular binding proteins.

II. MATERIALS* AND METHODS

A. Buffers

The assay buffer employed was 50 mM Tris-HCl, pH 7.4, containing theophylline (8 mM) and 2-mercaptoethanol (6 mM).

The medium (Medium A) used for preparation of binding reagent from a number of investigated tissues comprised 0.25 M sucrose, 25 mM potassium chloride, 5 mM magnesium chloride, and 50 mM-Tris-HCl buffer pH 7.4.

B. Tritium Labeled Cyclic AMP

This material was purchased from either New England Nuclear (NEN) (6072 Dreieichenbahn/Frankfurt a.M., Ger.), or the Radiochemical Centre (RCC) (Amersham, Eng.). The tracer currently in use has a specific activity of 24.1 C/mmole (NEN) or 25 C/mmole (RCC). This material, which is delivered in 50% ethanol-water, is diluted to a concentration of 5 μC/ml in 50% ethanol-water and stored in 200 μl portions at $-20°C$ until use. One aliquot is taken to dryness and resuspended in assay buffer for each assay run (approximately 6 pmoles/ml for 300 μl assays and approximately 1–2 pmoles/ml for 50 μl assays).

C. Binding Reagent

We have predominantly used bovine adrenal glands as the source of binding reagent and it is this preparation that will be described. It appears

*Unless otherwise stated all chemicals were purchased from B.D.H. Ltd. (Poole, Dorset, Eng.).

that a number of tissues from various animal species can be used although the conditions of the assay (e.g., pH) may need to be varied.

The bovine tissues (obtained from an abbatoir) were transported on ice to a cold room. They were stripped of fat and the relevant part (e.g., adrenal cortex) homogenized in a mixer-emulsifier (Silverson Machines Ltd. [London]) with 1.5 vol of Medium A. The supernatant obtained after centrifugation of the homogenate at 2,000 × g for 5 min was normally respun at 5,000 × g for 15 min. The centrifugal force applied has been found not to be critical. Thus the first centrifugation can be between 1,250 and 2,000 × g, and the second between 5,000 and 15,000 × g and probably higher. The final supernatant was aliquoted into 0.5 and 1 ml fractions and stored at − 20°C, a separate portion being thawed and diluted with the assay buffer for each assay. The complete procedure took approximately 3 hr and yielded between 100 and 300 ml of binding reagent (i.e., enough for about 30,000 to 100,000 assay tubes).

Binding protein dilution curves were set up for each new preparation in order to ascertain the dilution to be used in subsequent assays. (The dilution to be employed has usually been determined in the light of the optimization protocol as outlined in the theoretical part of this paper.)

D. Standard Cyclic AMP

This material (Sigma Chemical Co., London) was serially diluted for each assay to yield concentrations ranging between 5–160 pmoles/ml for assays of 300 μl final volume and between 1.25–40 pmoles/ml for assays in 50 μl final volume.

E. Assay Procedure

The assay was normally carried out in round bottom tubes 50 mm × 9.75 mm. An assay comprising 100 to 200 samples can be run conveniently. The protocol shown in Table 1 was followed for the assay for final incubation volumes of either 300 μl or 50 μl.

The incubation tubes are subsequently held at 4°C for at least 1.5 hr. During this time aliquots may be taken into scintillator in order to determine the total radioactivity in each tube. (This is essential when, as a minor variation of the protocol, [³H]-cAMP has been initially added to the biological tissue or fluid under test in order to monitor extraction recovery; this step may also be considered desirable in other circumstances.) More tubes, usually separate, are set up to determine the standard amount of radioactivity introduced into all incubation mixtures. It is also necessary to set up "control" tubes to check the efficiency of the separation step and "nonspecific" binding in the assay. In these control tubes the binding reagent is replaced by 100 μl (20 μl) buffer. An additional control occasionally used comprises all the reactants (i.e.,

TABLE 1. *Assay protocol*

	Standards (μl)	Samples (μl)
[^{3}H]-cyclic AMP	50 (10)	50 (10)[a]
standard cyclic AMP (or sample)	50 (20)	50 (20)
diluted binding reagent	100 (20)	100 (20)
buffer	100 (—)	100 (—)
total volume	300 (50)	300 (50)

[a] This is replaced by the relevant volume of buffer if the tracer cyclic AMP has been added to the sample for the purpose of ascertaining extraction recovery.

^{3}H cyclic AMP, binding reagent, and buffer) plus a gross excess of unlabeled cyclic AMP.

After the incubation period, separation of the free from bound moieties was effected by the addition to the tubes of 100 μl of a stirred suspension of 10 mg of charcoal (Norit GSX; Norit Clydesdale [Glasgow, Scot.])* in assay buffer containing 2% bovine serum albumin. The tubes were then briefly agitated (Rotamixer, Chemlab Instruments [Ilford, Eng.]) and centrifuged (1,200 × g for 15 min at 4°C)†. A 100-μl aliquot of the supernatant was subsequently taken into liquid scintillator vials for the estimation of radioactivity. Calibration curves were plotted from the standard data and the amounts of cyclic AMP in unknown samples are determined by reference to such curves (see Results). It is advisable that each sample be run in replicate and/or at different dilutions.

F. Sample Preparation

Tissue samples were prepared as described previously (Brown et al., 1971). However, most methods of tissue preparation would be suitable. Urine samples were simply diluted to yield an equivalent volume in each tube of between 1 μl and 0.1 μl (or less) depending on the anticipated nucleotide concentration. (Following infusions of parathyroid hormone to normal subjects as little as 0.1 or 0.05 μl of urine are sufficient for assay.)

The assay of plasma samples has posed a greater problem since great care must be taken to ensure that blood is rapidly centrifuged at low tempera-

---

*In this assay, as set up in our laboratory, the use of Norit GSX resulted in the most consistently reproducible results with low (approximately 0.3%) nonspecific binding. However, other charcoals, notably Norit Ultra, were found to be usable, whereas Norit A and Norit OL resulted in high nonspecific binding (5–11%).

†Centrifugation at room temperature resulted in response curves that were slightly displaced with respect to the normal curves,

ture and that the resulting plasma is immediately diluted into the (theophylline containing) assay buffer. This procedure is necessary to limit the extent of enzymatic attack by cyclic nucleotide phosphodiesterase present in blood plasma. However, we have found that unextracted plasma can be reliably assayed directly by this method only if the standards are set up in cyclic nucleotide-free plasma from the same subject. For reasons that are not immediately apparent, different plasmas exert slightly different nonspecific effects on the response curve.

III. RESULTS

A. Standard Preparation

A typical standard curve plotted in terms of the percentage of radioactivity bound against the amount of nucleotide is shown in Fig. 1. A transform (Ekins and Newman, 1970) of this data is shown in the lower curve of Fig. 2. In this plot the response metameter $\log (R-R_0)$ is plotted against log nucleotide concentration where R is the observed free/bound ratio and R_0 is the corresponding ratio at zero nucleotide concentration. This presentation is mathematically equivalent to the logit transform, used by Rodbard and Lewald (1970), and usually results in straight line plots which are particularly amenable to computer analysis. The upper line in Fig. 2 depicts data from an assay set up under near optimal conditions for high sensitivity and shows a detection limit of approximately 30 fmoles/tube.

Figure 3 demonstrates that a detection limit of less than 10 fmole is attainable using incubation volumes of 50 μl. This value is of the same order as that predicted theoretically. The investigation of the specificity charactcristics has been described previously (Brown et al., 1971). The most important competitors are cyclic IMP and cyclic GMP which react with relative potencies of approximately 7% and 0.5% under standard conditions. (It should be noted that the relative potency of competing substances in a saturation assay system can lie between unity and the ratio of the equilibrium constants, the exact value depending solely on the distribution ratio of radioactive label between bound and free moieties (Ekins, Newman, and O'Riordan, 1968; Rodbard and Lewald, 1970).) It has been shown elsewhere (Brown et al., 1971) that the assay method described here is virtually free of "blank" effects.

B. Cyclic AMP Levels in Biological Material

Urinary cyclic AMP excretion was measured in 16 normal subjects; all the values fell in the range 1.65–5.99 μmoles/24 hr (mean 3.58 $\pm$ 0.37 SEM). In a study related to the urinary excretion of cyclic AMP in manic–depressive psychosis (Brown, Salway, Albano, Hullin, and Ekins, *in press*) successive 24-hr specimens were assayed from patients for up to three months.

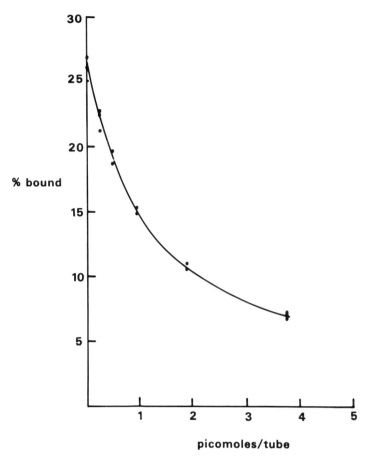

FIG. 1. Standard response curve for cyclic AMP: % of total tracer bound as a function of inactive cyclic AMP concentration.

The cyclic AMP levels measured by this method in rat tissue and in human plasma and urine are shown in Table 2. The concentrations of cyclic AMP are similar to those reported by others using a variety of techniques but are a little lower than those reported by Steiner and co-workers (Steiner, Parker and Kipnis, 1970).

IV. THEORETICAL ASPECTS

In this section we shall concern ourselves solely with the problem of maximizing assay sensitivity. It should be emphasized that this is not invariably the prime objective of the assayist. In many circumstances (e.g., the determina-

TABLE 2.

Sample	Treatment	Cyclic AMP Levels (pmoles/mg ± SEM)
rat liver	control	0.795 ± 0.052
	fasted	1.032 ± 0.074
	fasted + carbohydrate diet	0.771 ± 0.063
	fasted + fat diet	0.987 ± 0.074
	diabetic	0.867 ± 0.043
	diabetic and insulin	0.707 ± 0.066
rat liver	control (30 mg)	0.71
	control (15 mg)	0.72
		ng/incubation tube
rat adrenal[a] cells	ACTH stimulated	1.26
(zona fasciculata)	ACTH stimulated + phosphodiesterase[b]	0.08
		pmoles/μl
human urine	M.W.	1.08
	M.W.	1.12
	M.W. — Dowex-50w × 8 chromatography[c]	1.20
	M.W. — + phosphodiesterase[b]	0.03
	D.N.	5.14
	D.N.	5.26
	D.N. — + phosphodiesterase[b]	0.12
	J.W.	1.46
	J.W. — Ba/Zn precipitation[d]	1.55
	M.B.	1.22
	M.B. — Ba/Zn precipitation[d]	1.16
		pmoles/ml
normal human plasma		10–20

[a] Isolated rat adrenal zona fasciculata cells stimulated *in vitro* with ACTH (12.5 mU/ml)

[b] Treatment with beef heart cyclic nucleotide phosphodiesterase (Boehringer) for 15 min at 30°C.

[c] Dowex 50W × 8 washed with 0.1 N HCl; cyclic AMP eluted with deionized water. Results corrected for recovery (approx. 60%)

[d] Urine treated with barium hydroxide (0.3 N) and zinc sulfate (5%), subjected to centrifugation, diluted with buffer and assayed.

tion of urinary cyclic AMP) nucleotide levels are relatively high, and an assay designed to maximize the precision of measurement of concentrations in this range must employ entirely different concentrations of reagents from those used to maximize assay sensitivity. On the other hand it is often preferable, even in these circumstances, to design an assay for high sensitivity, and to rely on dilution of the biological fluid under test to bring the cyclic AMP concentration to within the useful range of the assay system. In this way, nonspecific effects (due to electrolytes, urea, etc.) which might otherwise invalidate assay

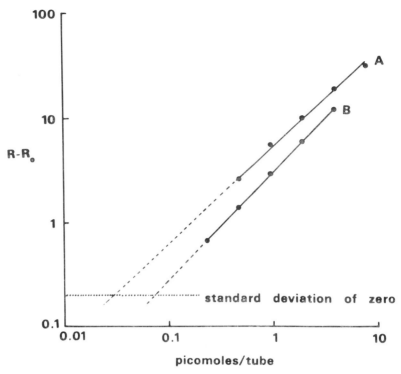

FIG. 2. Standard response curves relating the response metameter log (R-R_0) to log cAMP concentration. Curve A relates to an assay set up under near-optimal conditions, and yields a lower detection limit than curve B, the latter representing an assay suboptimal with respect to sensitivity. (The detection limit is represented in each case by the intercept of the response curve with the broken line, which indicates the standard deviation of the zero response.

results can be diluted out. A second reason for employing an assay system which is suboptimal with respect to sensitivity is that the experimenter may set a high priority on achieving a linear response curve. Reagent concentrations which achieve this end depend upon the particular coordinate system used (i.e., on the dose and response metameters selected), but in general will not represent a system which yields highest sensitivity. This approach is illustrated by Gilman's published method which involves the use of sufficient tracer cAMP to "saturate" the specific binding protein. Under these circumstances it is predictable that a doubling of the concentration of cyclic AMP in the system will halve the amount of tracer bound with the consequence that the response curve, plotted in a logarithmic coordinate system, will be linear. However, by adopting this approach Gilman knowingly (*personal communication*) sacrifices assay sensitivity.

There are nevertheless circumstances in which highest assay sensitivity

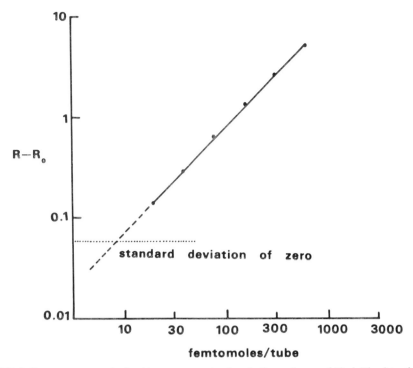

FIG. 3. Response curve obtained in an assay using incubation volumes of 50 μl. The detection limit in this assay is approximately 10 fmoles/tube.

is a prime requirement. The choice of reagent concentrations to achieve this objective has been discussed in detail elsewhere (Ekins et al., 1968; Ekins and Newman, 1970) and only the basic concepts and conclusions will be reviewed here. In general we may define an assay system as most sensitive when the detection limit is reduced to its smallest possible value. The detection limit is dependent on two parameters: the slope of the response curve at zero unlabeled cyclic AMP concentration (i.e., with only labeled cyclic AMP present), and the total error incurred in the measurement of the response metameter at this point. The latter, in turn, is made up of two independent components, the statistical error of counting, and the experimental error stemming from the pipeting of reagents, misclassification of bound and free nucleotide in the separation procedure, etc. Each of these factors (the response curve slope, and the two components of error) are dependent on the reagent concentrations used; hence the choice of concentrations to minimize the detection limit must represent an optimal compromise between the three parameters involved.

We shall simplify the situation by making two assumptions that are essentially appropriate in assays of cyclic AMP: (1) that only the bound, radioactive fraction is counted, and (2) that the relative error (ε) in the measurement of the response metameter (i.e., the percentage or fraction of total activity bound, r) is constant for all values of the response. Under these circumstances it may be shown (Ekins et al., 1970) that the optimal concentration of tracer (p^*_{opt}) and binding protein or antibody (q_{opt}) are given by:

$$q_{opt} = \frac{\left(\dfrac{1}{r_0} - 1\right)^2 + 2\left(\dfrac{1}{r_0} - 1\right) + 2}{K\left(\dfrac{1}{r_0} - 1\right)^3} \qquad \dots(1)$$

$$p^*_{opt} = \left(\frac{q_{opt}}{r_0} - \frac{1}{K(1 - r_0)}\right) \qquad \dots(2)$$

and

$$\left(\frac{1}{r_0} - 1\right)^4 - 2\left(\frac{1}{r_0} - 1\right)^3 - \frac{8\,\varepsilon^2\,SVT}{K}\left(\frac{1}{r_0} - 1\right) - \frac{8\varepsilon^2 SVT}{K} = 0 \quad \dots(3)$$

where K = equilibrium constant of the reaction

r_0 = the zero response using optimum reagent concentrations

S = tracer specific activity

V = volume of incubation mixture counted

T = counting time per sample

These propositions are summarized in Fig. 4 which illustrates the relationship between p^*_{opt} and q_{opt} and the product $\varepsilon\sqrt{SVT/K}$. Provided an estimate can be made of each of the variables in this product, it is a simple matter to compute the optimal concentrations of reagents which yield highest possible assay sensitivity.

The analysis also enables us to calculate the theoretical detection limit yielded using optimal reagent concentrations. The limit is given by:

$$\Delta p_{min} =$$

$$\frac{1}{r_0\sqrt{SVTK}}\left(Kq_{opt} + \left(\frac{r_0}{1 - r_0}\right)^2\right)\sqrt{\frac{\varepsilon^2 SVT}{K} + \frac{\left(\dfrac{1}{r_0} - 1\right)}{\left(\dfrac{1}{r_0} - 1\right)Kq_{opt} - 1}}$$

Figure 5 shows the variation in the calculated value as a function of the product SVT for different values of ε. The value of K assumed in the calculations was 1.66×10^9 L/M as derived from a Scatchard plot of a set of typical assay results using the adrenal binding protein. Figure 5 illustrates some fundamental conclusions relating to the effect of variation of certain assay parameters on sensitivity. Thus assuming that the incubation volume and counting time are

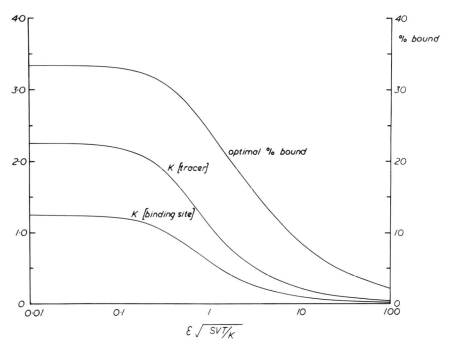

FIG. 4. Optimal concentrations of tracer and binding sites to yield maximum assay sensitivity plotted as a function of the product $\varepsilon\sqrt{SVT/K}$. The percentage of tracer cyclic AMP bound using these concentrations of reagents (at zero unlabeled nucleotide concentration) is also shown.

held constant, an increase in tracer specific activity will result in a decrease in the assay detection limit, i.e., an improvement in assay sensitivity. However, for any given value of ε, the enhancement in sensitivity stemming from an increase in tracer specific activity will progressively lessen, the value of the detection limit converging on a limiting value (ε/K) corresponding to tracer of infinitely high specific activity. This implies, of course, that experimental as opposed to counting errors progressively predominate in defining the value of the detection limit.

Reduction of the incubation volume increases the detection limit when the latter is expressed (as in Fig. 5) in terms of concentration; however, the detection limit expressed as the total weight of cAMP present in each assay tube (i.e., the product of concentration and incubation volume) will usually decrease when the incubation volume is decreased. This implies that a reduction in incubation volume will normally decrease the sensitivity of the system in terms of concentration but increase assay sensitivity to absolute amounts of nucleotide.

The value of these and similar conclusions which emerge from examination

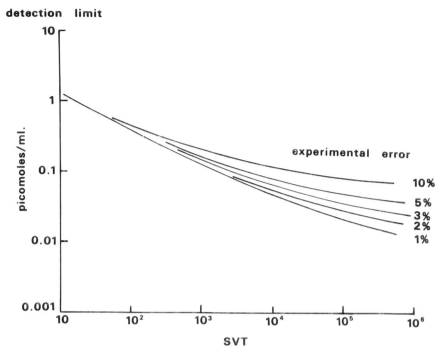

FIG. 5. The assay detection limit plotted as a function of the product SVT for different values of the experimental error (ε). This set of curves has been calculated assuming an effective equilibrium constant of 1.66×10^9 L/M as displayed by the adrenal binding protein in the system.

of Fig. 5 lie in the guidance they offer the experimenter in the area of assay design. In particular, the analysis permits the comparison of antibody and cellular binding protein methods from the point of view of the potential sensitivities attainable with the two classes of reagent.

Unfortunately it is not possible, on the basis of the published data of Steiner et al. (1970) to deduce a firm value for the equilibrium value displayed by the antiserum raised by these workers to cAMP. However, a reasonable estimate (assuming that the iodine-labeled derivative reacts with antibody with an energy identical to that of the unlabeled compound) is that the antibody reaction is characterized by a K value of approximately 3×10^8 L/M and this value will be assumed for the purposes of discussion in this paper. Figure 6 shows the set of curves corresponding to those shown in Fig. 5, relating the detection limit to the product STV. Comparison of the two sets of curves reveals that, for any given values of ε and the product SVT, the binding protein assay will yield higher sensitivity. However, considerably higher specific activities are attainable using the ^{125}I labeled nucleotide derivative as competitor. Steiner et al. (1970) claim specific activities of greater than 150 C/mmole,

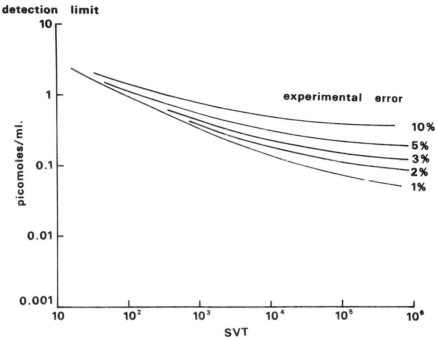

FIG. 6. The assay detection limit plotted as in Fig. 5 for an antibody displaying an equilibrium constant of 3×10^8 L/M. (S: cpm/pmole; V: ml; T: min.).

representing a factorial advantage of about ten over the tritiated nucleotide, and even higher specific activity iodinated material is, in principle, obtainable. This implies that the comparison must be made assuming at least a tenfold difference in the product STV, all other factors being held constant. Under these circumstances it is demonstrable that, in certain situations, the antibody/ ^{125}I-labeled derivative method will yield higher assay sensitivity. Thus if we assume a counting time (T) per sample of 1 min, an incubation volume of 5 μl and approximate specific activities of 10^4 cpm/pmole and 5×10^5 cpm/pmole for the tritiated and iodine-labeled compounds respectively, then, assuming an experimental error in the measurement of the response of the order of 3%, the antibody method will yield a lower detection limit (0.3 pmole/ml) than that relying on adrenal binding protein (0.6 pmole/ml). Conversely, for incubation volumes of 500 μl, the binding protein method is the more sensitive.

V. DISCUSSION

Recently a number of sensitive saturation assays for cyclic AMP have been developed. Steiner et al. (1969) used the globulin fraction of rabbit antisera raised in response to a succinyl-cyclic AMP-protein conjugate as a saturable reagent. The tracer ligand used to reflect the extent of competitive cyclic AMP binding was a radio iodinated tyrosine derivative of succinyl cyclic AMP. The specificity of at least one of the antisera raised by these workers was such that no purification of cyclic AMP containing extracts was necessary, only 2',3'-cyclic AMP showing slight cross-reaction. Likewise, the method described here, which depends on the specificity of a binding reagent prepared from bovine adrenal cortices, apparently obviates the need for purification of cyclic AMP-containing tissue extracts.

We have not as yet determined the extent to which the slight cross-reactivity observed with the other cyclic nucleotides reflects the presence in them of cyclic AMP contamination in addition to genuine competitive reactions. If it is shown that cross reacting nucleotides are likely to be present in the system at concentrations giving rise to interference it is clearly necessary to effect their removal—usually by means of a chromatographic separation of the unknown sample before assay. It is notable that 3',5'-cyclic inosine monophosphate, which shows the greatest cross-reactivity in this technique has been reported (Gill and Garren, 1970) to be the nucleotide of highest potency relative to cyclic AMP in the activation of the protein kinase of the bovine adrenal cortex.

The sensitivity attainable in a saturation method is dependent on a number of factors: on the affinity constant displayed by the binding reagent for its specific ligand, on the specific activity of the tracer available and on the experimental errors inherent in the manipulative procedures involved. In general, errors incurred in the determination of the response metameter do not greatly differ from one laboratory to another and cannot in practice be reduced below reasonable limits. In consequence, an increase in assay sensitivity is likely to arise only following a successful search for binding proteins demonstrating higher affinity constants. Other preparations with specific binding properties in terms of cyclic AMP have been described (Gilman, 1970; Walton and Garren, 1970; Cheung, 1970; Solomon and Schramm, 1970); it is conceivable that some will demonstrate a greater reaction energy than that characterizing the bovine adrenal cortex preparation and hence lead to greater sensitivity than we have attained. On the basis of published evidence the equilibrium constant displayed by muscle binding protein (Gilman, 1970) is closely comparable to that of the adrenal protein, and it is therefore unlikely that the currently available protein-binding methods differ markedly in sensitivity when set up under optimal conditions. However, some improvement in sensitivity using these methods is potentially attainable in certain circumstances by employing ^{32}P-labeled cyclic AMP of higher effective specific activity than the tritium-labeled reagent currently available, though with some increase in cost. In accordance with the theoretical predictions outlined in this paper, we have in practice attained sensitivities of the order of 75–100 fmoles/ml of incubation mixture using 300 μl incubation mixtures, and of 10 fmoles/tube using incubation volumes of 50 μl. These values have been achieved using tritiated material of 25 C/mmole.

The method described in this paper appears to present a number of useful features. Bovine adrenal glands are readily available. The preparation of the binding reagent is straightforward and may be readily accomplished in a morning or afternoon. Expertize in protein separation is not required. The crudely purified binding reagent is stable and may be stored at low temperature in amounts sufficient to provide for a large number of assays. The separation procedure to sequester free and bound moieties is simple and cheap, and numbers of tubes of 200 or greater can be conveniently run in a single assay. The method yields a sensitivity not as yet surpassed in other published assay techniques; moreover, by employing the transform described in this paper, response curves have normally been linear even in assays set up for maximum sensitivity. Nevertheless, the potential advantages of the antibody techniques should be emphasized. Although the method described by Steiner et al. (1970) has not in practice yielded the sensitivity for cyclic AMP that we have attained using adrenal binding protein (and is unlikely to do so except in special circumstances, if our own estimates of the avidity of available antisera are correct), further attempts may yield antisera of significantly higher reaction energy than naturally occurring binding proteins. Coupled to the possibility offered by antibody methods of using labeled nucleotide derivatives of very high specific activity, it is clear that in certain situations, the sensitivities that they will then yield will be of considerable value. Such circumstances will arise for example, in experiments with small numbers of isolated cells in which the absolute quantities of nucleotide synthesized in response to various stimuli are likely to be extremely small. Moreover, antibody methods are clearly applicable to cyclic nucleotides for which no naturally occurring binding proteins of adequate stability have been shown to exist.

ACKNOWLEDGMENTS

The authors thank Dr. P. McLean and Dr. K. Gumaa of the Courtauld Institute, The Middlesex Hospital Medical School, for their valuable assistance and expertise in preparing some of the tissue extracts. We are indebted to Professor J. F. Tait, Department of Physics, The Middlesex Hospital Medical School, for his encouragement and to Dr. D. V. Maudsley of the same department for helpful discussions. The financial assistance of the Medical Research Council is gratefully acknowledged.

REFERENCES

Brown, B. L., Ekins, R. P., and Tampion, W. (1970): The assay of adenosine 3′,5′-cyclic mono-phosphate by saturation analysis. *Biochemical Journal*, 120, 8p.

Brown, B. L., Albano, J. D. M., Ekins, R. P., Sgherzi, A. M., and Tampion, W. (1971): A simple and sensitive saturation assay method for the measurement of adenosine 3′,5′-cyclic monophosphate. *Biochemical Journal*, 121: 561–562.

Brown, B. L., Salway, J. G., Albano, J. D. M., Hullin, R. P., and Ekins, R. P.: Urinary excretion of cyclic AMP and manic-depressive psychosis. *British Journal of Psychiatry, in press*.

Cheung, W. Y. (1970): Adenosine 3′,5′-monophosphate: Demonstration of a binding site specific for the cyclic nucleotide. *Life Sciences*, 9:861–868.

Ekins, R. P., Newman, G. B., and O'Riordan (1968): Theoretical aspects of saturation and radio-immunoassay. In: *Radioisotopes in Medicine: In Vitro Studies*, edited by R. L. Hayes, F. A. Goswitz, and B. E. P. Murphy. Oak Ridge, Tennessee.

Ekins, R. P., and Newman, G. B. (1970): Theoretical aspects of saturation analysis. In: *Karolynska Symposia on Research Methods in Reproductive Endocrinology*, No. 2. edited by Diczfalusy. Karolinska Institutet, Stockholm.

Gill, G. N., and Garren, L. D. (1970): A cyclic-3′,5′-adenosine monophosphate dependent protein kinase from the adrenal cortex: Comparison with a cyclic AMP binding protein. *Biochemical and Biophysical Research Communications*, 39:335–343.

Gilman, A. G. (1970): A protein binding assay for adenosine 3′,5′-cyclic monophosphate. *Proceedings of the National Academy of Science*, 67:305–312.

Murad, F., Manganiello, V., and Vaughan, M. (1971): A simple, sensitive protein-binding assay for guanosine 3′,5′-monophosphate. *Proceedings of the National Academy of Science*, 68:731–739.

Rodbard, D., and Lewald, J. E. (1970): Computer analysis of radioimmunoassay and competitive protein binding assay data. In: *Karolinska Symposia on Research Methods in Reproductive Endocrinology*, No. 2, edited by Diczfalusy. Karolinska Institutet, Stockholm.

Solomon, Y., and Schramm, M. (1970): A specific binding site for 3′,5′-cyclic AMP in rat parotid microsomes. *Biochemical and Biophysical Research Communications*, 38, 106–111.

Steiner, A. L., Kipnis, D. M., Utiger, R., and Parker, C. W. (1969): Radioimmunoassay for the measurement of adenosine 3′,5′-cyclic phosphate. *Proceedings of the National Academy of Science*, 64: 367–373.

Steiner, A. L., Parker, C. W., and Kipnis, D. M. (1970): In: *Role of Cyclic AMP in Cell Function*, edited by P. Greengard and E. Costa. Raven Press, New York.

Walton, G. M., and Garren, L. D. (1970): An assay for adenosine 3′,5′-cyclic monophosphate based on the association of the nucleotide with a partially purified binding protein. *Biochemistry*, 9:4223–4229.

Advances in Cyclic Nucleotide Research, Vol. 2
Raven Press, New York © 1972

An Assay Method for Cyclic AMP and Cyclic GMP Based Upon Their Abilities To Activate Cyclic AMP-Dependent and Cyclic GMP-Dependent Protein Kinases

Jyh-Fa Kuo and Paul Greengard

*Department of Pharmacology, Yale University School of Medicine,
New Haven, Connecticut 06510*

I. PRINCIPLE

The assay method is based upon the ability of low concentrations of cyclic AMP and cyclic GMP to activate cyclic AMP-dependent and cyclic GMP-dependent protein kinases, respectively, which catalyze the phosphorylation of protein substrates (e.g., histone) by ATP. The reactions are depicted as follows:

$$\text{Histone} + [\gamma\text{-}^{32}\text{P}]\text{ATP} \xrightarrow[\text{cyclic AMP, Mg}^{++}]{\substack{\text{cyclic AMP-dependent} \\ \text{protein kinase}}} \text{Histone-}^{32}\text{P} + \text{ADP} \quad (1)$$

$$\text{Histone} + [\gamma\text{-}^{32}\text{P}]\text{ATP} \xrightarrow[\text{cyclic GMP, Mg}^{++}]{\substack{\text{cyclic GMP-dependent} \\ \text{protein kinase}}} \text{Histone-}^{32}\text{P} + \text{ADP}. \quad (2)$$

Under the standard assay conditions, the extent of histone phosphorylation is directly proportional to the amounts of cyclic AMP (reaction 1) or cyclic GMP (reaction 2) present in the incubation tubes, thus providing a direct, sensitive, and specific method for the measurement of each of these cyclic nucleotides in tissues and body fluids.

II. ANALYTICAL REAGENTS

A. Preparation of Cyclic AMP-Dependent Protein Kinase

In our laboratory we have isolated, studied, and compared cyclic AMP-dependent protein kinases from about 40 different sources (Kuo and Greengard, 1969; Kuo, Krueger, Sanes, and Greengard, 1970; Kuo, Wyatt, and Greengard, 1971). Of these, the enzyme from bovine heart has proven to be the most suitable for assaying cyclic AMP, because (1) it can be readily prepared from bovine heart with a good yield, (2) it has a high specific activity, and (3) it has a low K_a, i.e., high affinity, for cyclic AMP.

The procedure for the preparation of cyclic AMP-dependent protein kinase from bovine heart is similar to those described elsewhere for various tissues (Walsh, Perkins, and Krebs, 1968; Miyamoto, Kuo, and Greengard, 1969; Kuo et al., 1970). All steps used in the preparation of the enzyme are carried out at 4° or in ice. A bovine heart (about 1 kg, obtained either fresh from a local slaughterhouse or frozen from Pel-Freeze Biologicals, Inc.) is cut into small pieces, and homogenized with three volumes of neutral 4 mM ethylenediaminetetraacetic acid (EDTA) solution for 2 min in a Waring Blendor. The homogenate is centrifuged at 27,000 × g for 20 min. The supernatant solution is adjusted to pH 5.0 by the dropwise addition of 1 M acetic acid with stirring. After waiting 10 min, the precipitate is removed by centrifugation at 27,000 × g for 30 min. The pH of the clear supernatant is then readjusted to 6.5 with 1 M potassium phosphate buffer (pH 7.2). All buffers used in succeeding steps of the purification contain 2 mM EDTA.

Protein kinase activity is precipitated from the neutralized supernatant solution by the addition of solid ammonium sulfate (32.5 g/100 ml). After stirring for 30 min, the precipitate is collected by centrifugation at 27,000 × g for 20 min, and dissolved in 120 ml of 5 mM potassium phosphate buffer (pH 7.0). The resulting solution is dialyzed overnight against 20 volumes of the same buffer with two changes of buffer. After dialysis, the solution is centrifuged at 27,000 × g for 30 min, and the precipitate is discarded.

The enzyme solution is applied to a column (5 × 25 cm) of DEAE-cellulose that has been equilibrated with 5 mM potassium phosphate buffer (pH 7.0). After the enzyme is applied, the column is washed with two bed volumes of 0.05 M potassium phosphate buffer (pH 7.0); 0.3 M phosphate buffer (pH 7.0) is then applied to elute the enzyme. The active fractions are pooled and dialyzed overnight against 20 volumes of 5 mM potassium phosphate buffer (pH 7.0) with two changes of buffer. The dialyzed enzyme solution is used for assaying cyclic AMP.

B. Preparation of Cyclic GMP-Dependent Protein Kinase

The procedure for preparing cyclic GMP-dependent protein kinase is essentially the same as reported elsewhere (Kuo and Greengard, 1970a). About 300 g of tail muscle from three live lobsters is cut into small pieces and homogenized in a Waring Blendor with three volumes of neutral 4 mM EDTA for 2 min. The homogenate is centrifuged at $27,000 \times g$ for 20 min. The two subsequent steps, i.e., pH 5.0 precipitation and ammonium sulfate fractionation, are the same as described for cyclic AMP-dependent protein kinase. The ammonium sulfate fraction of lobster tail muscle contains both cyclic AMP-dependent and cyclic GMP-dependent protein kinase activities; it is desirable to separate these. This is accomplished by the following step. The dialyzed enzyme solution from the ammonium sulfate step is applied to a column (2.4 × 14 cm) of DEAE-cellulose equilibrated previously with 5 mM phosphate buffer at pH 7.0. Protein is eluted from the column by stepwise application of 100 ml each of 5 and 50 mM phosphate buffer at pH 7.0. Cyclic GMP-dependent protein kinase activity is associated with the protein peak eluted by 5 mM phosphate buffer, and cyclic AMP-dependent protein kinase activity is found in the protein peak eluted by the 50 mM phosphate buffer. The cyclic GMP-dependent enzyme eluted from the column is used directly for assaying cyclic GMP.

We have recently found that fat body tissue of cecropia silkmoth pupae appears to contain cyclic GMP-dependent protein kinase almost exclusively (Kuo et al., 1971); i.e., it contains only barely detectable cyclic AMP-dependent enzyme activity. The silkmoth enzyme from the ammonium sulfate step of purification, without further chromatography on a DEAE-cellulose column, is found to be suitable as a reagent for cyclic GMP assay. This particular tissue, however, may not be readily available to most investigators.

C. Preparation of Biological Materials for Cyclic AMP and Cyclic GMP Assay

Method 1. Tissues or cells (about 5 mg to 2 g wet wt) are homogenized in glass homogenizers (or alternatively in a sonicator) in cold 5% trichloracetic acid. If the cyclic nucleotide levels are to be measured in body fluids, one-tenth volume of 50% trichloracetic acid is added with mixing. In most cases, eight-tenths of the total volume of tissue homogenate or body fluid is then removed and used for the measurement of cyclic GMP as described below. To the remaining two-tenths of the sample, a trace amount of cyclic AMP-8-^{3}H (about 1×10^{-13} mole; 10^3 cpm) is added in a volume of 5 μl for the purpose of determining the recovery of cyclic AMP, and the tube is mixed. After removal of the precipitate by centrifugation, the pH of the supernatant is neutralized with 1 M Tris. The cyclic AMP in the supernatant is first purified by the BaSO$_4$ method of Krishna, Weiss, and Brodie (1968), by adding 0.04 ml each of 5.0%

zinc sulfate and 2.6% barium hydroxide. Barium sulfate precipitates and is removed by centrifugation, and the entire supernatant solution (usually 0.2 ml) from each sample is loaded onto a small column (0.5×2.5 cm) of AG 50W-X8 (100–200 mesh, H^+-form, BioRad) packed in a short-tip disposable pipet, the resin having been previously washed with water. Cyclic AMP is collected in the third through fourth ml, the column being eluted with water. Overall recovery of the tissue cyclic AMP is usually between 65 and 70%. Aliquots (usually 0.02 to 0.2 ml) of the cyclic AMP fraction from the column are lyophilized in small tubes (1.3×10.0 cm) in which the cyclic nucleotide is to be assayed.

The remaining eight-tenths of the tissue homogenate is used for the purification of cyclic GMP (Kuo, Lee, Reyes, Walton, Donnelly, and Greengard, 1972). A 5 μl aliquot of cyclic GMP-^{3}H (about 5×10^{-13} mole; 10^3 cpm) is added to the tissue homogenate for the purpose of determining the recovery of tissue cyclic GMP. The precipitate is removed by centrifugation, and the supernatant solution (usually 0.5 to 1.5 ml) is charged onto an 0.5×2.0 cm column of AGl-X8 (200–400 mesh, formate form), the resin having been previously washed with water. The column is then washed with 6 ml of 0.5 N formic acid and the eluate, which contains cyclic AMP, is discarded. Cyclic GMP is then eluted from the column with 3 ml of 4 N formic acid and this eluate is lyophilized. The dried material is taken up in 0.2 ml of water and quantitatively spotted on a thin layer chromatographic plate (SilicAT TLC-7GF, Mallinckrodt). The plate is developed using the solvent system of Goldberg, Dietz, and O'Toole (1969): isopropyl alcohol–H_2O–2.9% NH_4OH (7:2:1, v/v/v). Cyclic GMP is then eluted with two 1.0-ml aliquots of absolute ethanol from the area on the plate corresponding to the spot of authentic compound. Overall recovery of tissue cyclic GMP is between 70 and 80%. Aliquots (usually 0.3 to 1.0 ml) of the alcoholic eluates of cyclic GMP are dried at 65°, *in vacuo*, in the test tubes (1.3×10.0 cm) in which the cyclic nucleotide is to be assayed.

Method 2. More recently we have modified the procedure for the separation and purification of cyclic AMP and cyclic GMP by including alumina oxide chromatography and omitting the barium sulfate and thin layer chromatography steps (Lee, Kuo, and Greengard, 1972). In this newer procedure, the entire trichloracetic acid tissue extract is centrifuged, and an aliquot (0.5 to 1.0 ml) of the neutralized supernatant is charged onto a column (0.5×8.0 cm) of AG50W-X8. Cyclic GMP is eluted from the column with 5 ml of 1 mM potassium phosphate buffer, pH 7.0. Cyclic AMP is subsequently eluted from the column with 5 ml of water. Aliquots of this eluate are lyophilized in small tubes and then assayed for cyclic AMP as described below. The 5-ml fraction eluted with phosphate buffer which contains cyclic GMP is loaded onto a column (0.5×2.0 cm) of neutral alumina oxide (Sigma). About one-fifth of the cyclic GMP appears in the 5 ml of effluent in this step. The remainder of the cyclic GMP is eluted by adding 2 ml of 50 mM Tris-Cl buffer, pH 9.0, to the alumina oxide column. The combined effluent (total 7 ml) containing the cyclic GMP is charged onto a column (0.5 × 2.0 cm) of AGl-X8, and the column is washed with 10 ml of 0.5 N formic acid. The cyclic GMP is then eluted from the column with 4 ml of 4 N formic acid. Aliquots of the eluate are lyophilized in small tubes and then assayed for cyclic GMP as described below. The overall recovery of tissue cyclic AMP in Method 2 is greater than 80% and that for cyclic GMP is about 60%. The results obtained with the two methods are comparable. Alumina oxide was reported earlier by White and Zenser (1971) to be useful in the purification of cyclic nucleotides.

D. Miscellaneous

1. $[\gamma\text{-}^{32}P]ATP$: In our laboratory, this is prepared from *ortho*-^{32}P according to the method of Post and Sen (1967). However, it can be obtained from commercial sources (ICN, New England Nuclear).

2. *Trichloracetic acid-sodium tungstate-sulfuric acid precipitating solution.* Slowly dissolve 12 g of NaOH pellets, with stirring, in 1 liter of 5% trichloracetic acid (the pH of the resultant solution should be 2) followed by dissolving 2.5 g of $NaWO_4 \cdot 2H_2O$. Readjust the pH to 2 with 100% trichloracetic acid, and finally add 2.0 ml of concentrated (36 N) sulfuric acid. The resultant solution remains clear for several months.

3. *Histone mixture.* Dissolve 20 mg of histone (mixture, calf thymus, Schwarz/Mann) in 9 ml of 0.01 N HCl. Neutralize the solution with 5 N sodium hydroxide and add water to make a final volume of 10 ml. The resultant histone solution should remain clear after neutralization.

III. ASSAY FOR CYCLIC NUCLEOTIDES

A. Cyclic AMP

For the assay (Kuo and Greengard, 1970b), the components below are added to the incubation tubes (some of which contain lyophilized sample) in the order given.

Solution	Volume (ml)
sodium acetate buffer, 0.1 M, pH 6.0	0.100
magnesium acetate (2 μmoles)	0.020
cyclic AMP standard (0–10 pmoles) or water	0.020
histone mixture (40 μg)	0.020
cyclic AMP-dependent protein kinase (50–200 units)	0.020
$[\gamma\text{-}^{32}P]ATP$ (1000 pmoles)	0.006

The tubes are kept in ice during the addition of the enzyme. One unit of enzyme is defined as that amount of enzyme which transfers 1 pmole of ^{32}P from $[\gamma\text{-}^{32}\text{P}]$ATP to histone in 5 min at 30° under the assay conditions. The reaction is commenced by the addition of radioactive ATP (containing about 1.5 to 2.2 × 10^6 cpm), which is conveniently delivered from a 0.1-ml Hamilton microsyringe (with a fused needle) attached to a repeating dispenser. The tubes are incubated for 5 min at 30°, with shaking, and the reaction is terminated by the addition of 2 ml of ice-cold trichloracetic acid-tungstate-sulfuric acid (hereinafter referred to as the precipitating solution). Then 0.2 ml of 0.6% bovine serum albumin is added as a carrier protein and the contents of the tube are mixed by vigorous addition of another 2 ml of the precipitating solution. The mixture is centrifuged, and the supernatant solution is removed by aspiration.

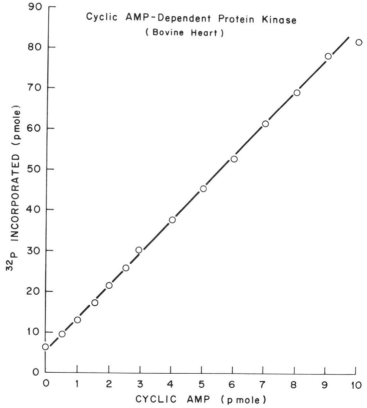

FIG. 1. Standard curve for the measurement of cyclic AMP with cyclic AMP-dependent protein kinase prepared from bovine heart. One pmole of ^{32}P incorporated represents about 2,000 cpm.

The precipitate is dissolved in 0.1 ml of 1 N NaOH, and 2 ml of the precipitating solution is added. The procedure of centrifuging, removing the supernatant, dissolving the precipitate in alkali, and reprecipitating the protein is repeated once more. The protein is finally collected by centrifugation and dissolved in 0.1 ml of 1 N NaOH, and the radioactivity is counted in 6 ml of scintillation fluid (made by dissolving 8 g of Omniflour, purchased from New England Nuclear, in 1 liter of toluene and 1 liter of ethylene glycol monoethyl ether). The amount of cyclic AMP present in the samples is determined from standard curves obtained by assaying known quantities of cyclic AMP.

Figure 1 shows a typical standard curve for cyclic AMP assay, and some of the results are presented in Table 1. In an earlier report (Kuo and Greengard, 1970*b*) we showed that the standard curve for cyclic AMP measurement had two distinct slopes when batches of histone mixture purchased from Mann were used. In more recent studies, we have found that a standard curve with a single slope, as illustrated in Fig. 1, is obtained when new batches of histone mixture purchased from Schwarz/Mann are used as the substrate.

TABLE 1. *Basal levels of cyclic AMP and cyclic GMP in some representative tissues and regulation by various hormones of their levels in incubated slices of rat ventricle, and rabbit cerebellum and cerebrum.*

Experiment	Tissue level (pmoles/mg protein)		A/G Ratio
	Cyclic AMP	Cyclic GMP	
rat cerebrum, untreated	12.7	0.20	63.5
rat cerebellum, untreated	6.3	4.29	1.5
rat lung, untreated	8.4	5.81	1.4
rat liver, untreated	2.7	0.14	19.3
calf atrium, untreated	7.8	0.38	20.5
calf ventricle, untreated	13.1	0.18	72.7
rat ventricle, control, 1 min	4.5 ± 0.4	0.32 ± 0.08	14.1
acetylcholine (0.3 μM), 1 min	4.3 ± 0.2	3.15 ± 0.12	1.4
isoproterenol (0.3 μM), 1 min	12.8 ± 0.9	0.29 ± 0.01	44.1
rabbit cerebellum, control, 5 min	14.4	1.50	9.6
histamine (1 μM), 5 min	59.2	1.35	43.9
norepinephrine (0.1 μM), 5 min	166.6	0.71	234.6
acetylcholine (0.5 μM), 2 min	10.4	2.57	4.1
rabbit cerebral cortex, control, 15 min	20.0	0.68	29.4
histamine (1 μM), 15 min	250.5	1.12	223.7
acetylcholine (0.5 μM), 2 min	24.7	1.22	20.2

The procedure for the preparation of heart slices was as described by Lee et al. (1971), and for brain slices as described by Kakiuchi and Rall (1968). The tissue slices were incubated in physiological salt solutions in the presence of the agents indicated. The values obtained with rat ventricular slices represent means ($\pm$ standard error) of assays on triplicate incubations. The other values represent means of triplicate assays on one or two tissue samples. (Modified from Kuo et al., 1972.)

B. Cyclic GMP

The procedure for assaying cyclic GMP is essentially the same as for cyclic AMP except that cyclic GMP samples or standards (up to 20 pmoles) and a cyclic GMP-dependent protein kinase (50–200 units), from either lobster tail muscle or silkmoth fat body, are substituted for cyclic AMP and cyclic AMP-dependent protein kinase, respectively (Kuo et al., 1972). Typical standard curves for the cyclic GMP assay are shown in Fig. 2, and some of the results, together with cyclic AMP values obtained from the same tissue samples, are presented in Table 1.

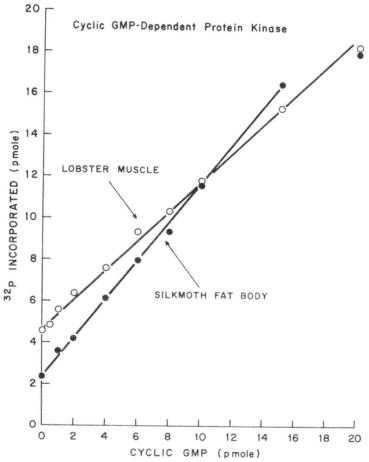

FIG. 2. Standard curves for the measurement of cyclic GMP with cyclic GMP-dependent protein kinase prepared from either lobster tail muscle or cecropia silkmoth pupal fat body. One pmole of ^{32}P incorporated represents about 2,000 cpm. (Modified from Kuo et al., 1972).

IV. CONCLUDING REMARKS

The limit of sensitivity of the procedure for assaying cyclic AMP with cyclic AMP-dependent protein kinase is about 0.3 pmole, and that for assaying cyclic GMP with cyclic GMP-dependent protein kinase is about 0.5 pmole. The preliminary purification of tissue cyclic nucleotides is found to separate effectively these two cyclic nucleotides from each other and remove any substances, such as ATP, originally present in the crude tissue extracts, which might interfere with the assay for cyclic AMP and cyclic GMP based upon their abilities to stimulate cyclic AMP-dependent and cyclic GMP-dependent protein kinase, respectively. The assay method described here has been shown to be useful for determining the levels of cyclic nucleotides in a wide variety of tissues under various experimental conditions.

The incubation system for measuring protein kinase activity has been modified from the one described earlier by Walsh et al. (1968), since we found that: (1) protein kinase is much more active in acetate buffer than in glycerol phosphate buffer, (2) histone is a much better substrate (at least 30 times more reactive) than casein, and (3) theophylline and sodium fluoride are unnecessary and therefore are omitted from the assay system. Trichloracetic acid-tungstate-sulfuric acid is employed for quantitative recovery of histone from the reaction mixture, since histone is partially soluble in 5% trichloracetic acid.

An assay method for the measurement of cyclic AMP based upon a principle similar to ours (Kuo and Greengard, 1970b) has been reported (Wastila, Stull, Mayer, and Walsh, 1971). These authors showed that they were able to measure cyclic AMP directly in crude tissue extracts, without any preliminary purification of the cyclic nucleotide. The interference with the protein kinase activity by materials present in the tissue extracts was apparently minimized by using high concentrations of casein (as protein substrate) and $[\gamma\text{-}^{32}P]ATP$ in the incubation system.

Acknowledgments. This work was supported by Grants HE-13305, NS-08440 and MH-17387 from the U.S. Public Health Service, by Grant GB-27510 from the National Science Foundation, and by Grant G-70-31 from the Life Insurance Medical Research Fund. One of us (J.F.K.) is recipient of a Research Career Development Award (1 K4 GM-50165) from the U.S. Public Health Service.

REFERENCES

Goldberg, N. D., Dietz, S. B., and O'Toole, A. G. (1969): Cyclic guanosine 3′,5′-monophosphate in mammalian tissues and urine. *Journal of Biological Chemistry*, 244:4458–4466.
Kakiuchi, S., and Rall, T. W. (1968): Studies on adenosine 3′,5′-monophosphate in rabbit cerebral cortex. *Molecular Pharmacology*, 4:379–385.

Krishna, G., Weiss, B., and Brodie, B. B. (1968): A simple, sensitive method for the assay of adenyl cyclase. *Journal of Pharmacology and Experimental Therapeutics*, 163:379–385.

Kuo, J. F., and Greengard, P. (1969): Cyclic nucleotide-dependent protein kinases. IV. Widespread occurrence of adenosine 3',5'-monophosphate-dependent protein kinase in various tissues and phyla of the animal kingdom. *Proceedings of the National Academy of Sciences*, 64:1349–1355.

Kuo, J. F., and Greengard, P. (1970a): Cyclic nucleotide-dependent protein kinases. VI. Isolation and partial purification of a protein kinase activated by guanosine 3',5'-monophosphate. *Journal of Biological Chemistry*, 245:2493–2498.

Kuo, J. F., and Greengard, P. (1970b): Cyclic nucleotide-dependent protein kinases. VIII. An assay method for the measurement of adenosine 3',5'-monophosphate in various tissues and a study of agents influencing its level in adipose cells. *Journal of Biological Chemistry*, 245:4067–4073.

Kuo, J. F., Krueger, B. K., Sanes, J. B., and Greengard, P. (1970): Cyclic nucleotide-dependent protein kinases. V. Preparation and properties of adenosine 3',5'-monophosphate-dependent protein kinase from various bovine tissues. *Biochimica et Biophysica Acta*, 212:79–91.

Kuo, J. F., Wyatt, G. R., and Greengard, P. (1971): Cyclic nucleotide-dependent protein kinases. IX. Partial purification and some properties of guanosine 3',5'-monophosphate-dependent and adenosine 3',5'-monophosphate-dependent protein kinases from various tissues and species of arthropoda. *Journal of Biological Chemistry*, 246:7159–7167.

Kuo, J. F., Lee, T. P., Reyes, P. L., Walton, K. G., Donnelly, T. E., and Greengard, P. (1972): Cyclic nucleotide-dependent protein kinases. X. An assay method for the measurement of guanosine 3',5'-monophosphate in various biological materials and a study of agents regulating its levels in heart and brain. *Journal of Biological Chemistry*, 247:16–22.

Lee, T. P., Kuo J. F., and Greengard, P. (1971): Regulation of myocardial cyclic AMP by iso-proterenol, glucagon and acetylcholine. *Biochemical and Biophysical Research Communications*, 45:991–997.

Lee, T. P., Kuo, J. F., and Greengard, P.: Regulation of ilial cyclic GMP and cyclic AMP by acetylcholine and catecholamines. (*in preparation*).

Miyamoto, E., Kuo, J. F., and Greengard, P. (1969): Cyclic nucleotide-dependent protein kinases. III. Purification and properties of adenosine 3',5'-monophosphate-dependent protein kinase from bovine brain. *Journal of Biological Chemistry*, 244:6395–6402.

Post, R. L., and Sen, A. K. (1967): Sodium and potassium stimulated ATPase. *In*: *Methods of Enzymology*, Vol. X, edited by R. W. Estabrook and M. E. Pullman, 773–775. Academic Press, New York.

Walsh, D. A., Perkins, J. P., and Krebs, E. G. (1968): An adenosine 3',5'-monophosphate-dependent protein kinase from rabbit skeletal muscle. *Journal of Biological Chemistry*, 243:3763–3765.

Wastila, W. B., Stull, J. T., Mayer, S. E., and Walsh, D. A. (1971): Measurement of cyclic 3',5'-adenosine monophosphate by the activation of skeletal muscle protein kinase. *Journal of Biological Chemistry*, 246:1996–2003.

White, A. A., and Zenser, T. V. (1971): Separation of cyclic 3',5'-nucleoside monophosphates from other nucleotides on alumina oxide columns. Application to the assay of adenyl cyclase and guanyl cyclase. *Analytical Biochemistry*, 41:372–396.

Advances in Cyclic Nucleotide Research, Vol. 2
Raven Press, New York © 1972

Radioimmunoassay for the Measurement of Cyclic Nucleotides

Alton L. Steiner, Robert E. Wehmann, Charles W. Parker, and David M. Kipnis

Department of Medicine, Albany Medical College, Albany, New York 12208, and the Divisions of Metabolism and Immunology, Department of Medicine, Washington University School of Medicine, St. Louis, Missouri

I. INTRODUCTION

We have previously described the development of radioimmunoassays for the measurement of cyclic 3',5' adenosine monophosphate (cAMP), cyclic 3',5' guanosine monophosphate (cGMP), cyclic 3',5' inosine monophosphate (cIMP), and cyclic 3',5' uridine monophosphate (cUMP) (Steiner, Parker, and Kipnis, 1970). The sensitivity and specificity of the cAMP and cGMP immunoassays permits the measurement of these cyclic nucleotides on small quantities of tissue without the need for chromatographic techniques. These assays are based upon the competition of the cyclic nucleotide with an isotopically labeled derivative for binding sites on specific antibody, and in principle are similar to the radioimmunoassay technique described by Yalow and Berson (1960) for the measurement of peptide hormones. The radioimmunoassay technique for the measurement of the cyclic nucleotides has been modified to allow simultaneous measurement of cAMP and cGMP in tissue extracts and body fluids.

II. PREPARATION OF REAGENTS FOR CYCLIC NUCLEOTIDE RADIOIMMUNOASSAYS

A. Synthesis of 2'-O-Succinyl Cyclic Nucleotide

The cyclic nucleotides were rendered immunogenic by conjugating a succinylated derivative of the cyclic nucleotide to protein. The cyclic nucleotides were succinylated at the 2'O position with succinic anhydride in anhydrous pyridine by a modification of the method of Falbriard, Posternak, and Sutherland (1967) for synthesizing monocarboxylic acid derivatives of cAMP, and the free carboxyl group of this derivative was then conjugated to protein. As described previously (Steiner et al., 1970), cAMP, cUMP, and cIMP were solubilized in pyridine by the addition of equimolar quantities of 4'-morpholine N,N'-dicyclohexylcarboxamidine, or triethylamine to form either the morpholinium or triethylammonium salt of the cyclic nucleotide. Partial solubilization of cGMP was achieved by forming the trioctylammonium salt in anhydrous pyridine. Succinic anhydride was added in excess to the suspension of cyclic nucleotide in pyridine, and the reaction mixture was stirred for 18 hr at room temperature. Unreacted succinic anhydride was converted to succinic acid by the addition of water, and the pyridine was removed by rotary evaporation at 40°C under reduced pressure. Succinyl cAMP (ScAMP) was purified by chromatography on a 1.5 cm × 44 cm column of Dowex-50 (H^+ form) using distilled H_2O as the eluent. Initially, the fractions containing ScAMP were pooled, lyophilized, and further purification accomplished by preparation of the barium salt. However, this purification step has been discarded since the product isolated from the Dowex-50 column fractionation routinely exhibited greater than 95% purity on thin layer chromatography. Following brief treatment with 0.1 N NaOH, the product quantitatively reverted to cAMP, confirming that the succinyl substitution was exclusively at the 2'O position (Falbriard et al., 1967). Succinyl cyclic GMP (ScGMP), succinyl cyclic IMP (ScIMP), and succinyl cyclic UMP (ScUMP) were purified by thin layer chromatography on cellulose using a solvent system butanol: glacial acetic acid:H_2O (12:3:5, v/v/v). In this system the succinylated cyclic nucleotides run ahead (R_f 0.42) of the unreacted cyclic nucleotide (R_f 0.30).

B. Preparation of Antigen

The preparation of antigen has been described previously (Steiner et al., 1970). Succinylated cyclic nucleotide was coupled to protein (i.e., human serum albumin, keyhole limpet hemocyanin) or poly-L-lysine polymers. ScAMP (10 mg) was reacted with 20 mg human serum albumin and 10 mg 1-ethyl-3-(3-dimethylaminopropyl)-carbodiimide-HCl (EDC) in aqueous solution at pH 5.5. After incubation of this mixture at 24°C for 16 hr in the dark, the conjugate was dialyzed against phosphosaline buffer (0.01 M sodium phosphate, 0.15 M sodium chloride, pH 7.4) for 48 hr. The dialyzed conjugate (ScAMP-albumin) exhibited an absorption maximum at 260 nm. On the basis of the spectrum of ScAMP-albumin and unconjugated human serum albumin, and assuming a molar extinction coefficient of 15,000 for ScAMP, the conjugate was estimated to contain an average of 5 to 6 cyclic AMP residues per albumin molecule. ScAMP was also coupled to poly-L-lysine and keyhole limpet hemocyanin using the same method, with similar results. ScUMP, ScIMP, and ScGMP were coupled to keyhole limpet hemocyanin by the same procedure outlined for ScAMP.

Rabbits were immunized by the injection into each footpad of 0.25 mg of the cyclic nucleotide protein conjugate which had been emulsified in complete Freund's adjuvant. Booster injections with 0.2 to 0.4 mg cyclic nucleotide protein conjugate into two footpads were given at 4 to 6 week intervals, and the animals bled 10 to 14 days later. Antibody in adequate titer was found in a majority of animals after one booster injection.

C. Synthesis of 2'-O-Succinyl cyclic Nucleotide Tyrosine Methyl Ester

A radioactive derivative of cAMP of high specific activity was synthesized by tyrosination of the succinylated cyclic nucleotide derivative and subsequent iodination of the tyrosine moiety. The synthesis of ScAMP-TME using N, N' dicyclohexyl carbodiimide as the coupling agent was described in a previous paper (Steiner, Kipnis, Utiger, and Parker, 1969). Because of the poor yield with that reaction, the mixed carboxylic-carbonic acid reaction (Greenstein and Winitz, 1961) using ethyl chloroformate was tried and found to give significantly better results. This reaction has been used for the synthesis of the 2'-O-succinyl cyclic nucleotide tyrosine methyl ester derivatives of cAMP, cGMP, cIMP, and cUMP. One equivalent (5 μmole) of the succinylated cyclic nucleotide was dissolved in 0.1 ml dimethylformamide (DMF) at 0°C with three equivalents of trioctylamine. All the succinylated cyclic nucleotides readily dissolved in DMF except for ScGMP, which remained as a fine suspension. One equivalent of ethyl chloroformate in DMF was added and the reaction allowed to proceed at 0°C for 15 min. Two equivalents of both tyrosine methyl ester hydrochloride and trioctylamine were then added in 0.1 ml DMF and the reaction continued at room temperature for an additional 2 to 3 hr with continuous stirring. The tyrosinated product was isolated by thin layer chromatography on cellulose with the solvent system butanol: glacial acetic acid: H$_2$O (12:3:5, v/v/v). The new nitrosonaphthol positive derivative (R_f 0.57) ran ahead of the succinylated cyclic nucleotide (R_f 0.42) and behind unreacted tyrosine methyl ester hydrochloride (R_f 0.65). The tyrosinated derivatives exhibited an absorption maximum in water identical to that of the parent cyclic nucleotide.

D. Preparation of Radioactive ^{125}I or ^{131}I-Succinyl Cyclic Nucleotide Tyrosine Methyl Ester

Succinyl cyclic nucleotide tyrosine methyl ester was iodinated with ^{125}I or ^{131}I by the method of Hunter and Greenwood (1962). Approximately 2 to 3 μg of the derivative (in 50 μl water) was added to 40 μl of 0.5 M phosphate buffer, pH 7.5. After the addition of 0.5 to 1.0 mC ^{125}I or ^{131}I, 50μl of a solution of chloramine-T (35 mg/10 ml 0.05 M phosphate buffer) was added and the reaction run for 45 sec. The iodine was then reduced by the addition of 100 μl of a solution of sodium metabisulfite (24 mg/10 ml 0.05 M phosphate buffer.)

The iodinated cyclic nucleotide derivatives were purified either by column chromatography on Sephadex G-10 or by thin layer chromatography on cellulose. The reaction mixture was applied to a 0.9 cm $\times$ 9 cm Sephadex G-10 column previously washed with 1 ml of 3% human serum albumin in phosphosaline buffer, pH 7.5, and eluted with phosphosaline buffer (flow rate 40 ml/hr). Three distinct peaks of radioactivity were found: peak 1 (void volume) has not been identified, peak 2 (9 to 12 ml) was free iodine, and peak 3 (22 to 32 ml) was ^{125}I-succinyl cyclic nucleotide. The iodinated tyrosine methyl ester derivatives isolated in peak 3 co-chromatographed with their respective uniodinated compounds on thin layer chromatography using the previously described solvent system. All iodinated compounds had a specific activity of > 150 C/mmole. The iodinated ligands were diluted in 0.05 M acetate buffer, pH 6.2 (3 to 4 $\times$ 10^6 cpm/ml) and stored as small aliquots at $-20°$C. The ^{125}I material retained full immunoreactivity for periods up to 2 months or longer, provided it was stored at $-20°$C in small aliquots and not subjected to repeated freezing and thawing. The ^{131}I derivatives were stable for 3 to 4 weeks.

III. PREPARATION OF TISSUES, BLOOD, AND URINE

Rat tissues were frozen *in situ* between stainless steel tongs cooled in liquid nitrogen, and were stored at $-80°$C. Frozen tissue (5 to 100 mg) was powdered at $-20°$C and homogenized at 4°C in 0.5 ml of 6% trichloroacetic acid. After centrifugation at 2,500 $\times$ g for 15 min, the supernatant fluid was removed and extracted three times with 5 ml of ethyl ether saturated with water. The extracted aqueous phase was evaporated under a stream of air and the residue dissolved in 0.05 M sodium acetate buffer, pH 6.2, and used directly in the immunoassay.

Blood was collected in heparinized tubes and centrifuged immediately at 2,500 $\times$ g for 5 min at 4°C. Extracts of plasma were prepared by adding equal volumes of plasma and 10% trichloroacetic acid and centrifuging for 15 min. The supernatant fraction was then treated in a manner identical to that of tissue extracts. Urine samples (2 to 10 μl) were added directly into the immunoassay.

IV. IMMUNOASSAY PROCEDURE

cAMP and cGMP immunoassays were performed in 0.05 M sodium acetate buffer, pH 6.2; cIMP and cUMP assays were carried out in 0.05 M imidazole buffer, pH 7.5. Each tube contained (in order of addition) 50 to

300 μl of cyclic nucleotide standard or unknown solution, 100 μl of antibody in appropriate buffer at a dilution sufficient to bind 35 to 55% of the labeled marker, 100 μl of the ^{125}I-labeled marker (approximately 15,000 cpm and representing < 0.01 pmoles ligand), and 100 μl containing 500 μg of rabbit gamma globulin as carrier, in a final volume of 600 μl. The most commonly employed method for separating bound and free ^{125}I ligand now used in our laboratory is ammonium sulfate precipitation. After 2 to 18 hr of incubation, 2.5 ml of 60% $(NH_4)_2SO_4$ solution was added. The tubes were centrifuged at 4°C for 15 min, and the precipitate was counted in the gamma spectrometer. All analyses were carried out in triplicate.

In the simultaneous assay of cAMP and cGMP, the immunoassay procedure was identical, except that specific cAMP and cGMP antibodies were added in the 100 μl antibody aliquot. ^{131}I-ScAMP-TME and ^{125}I-ScGMP-TME (approximately 15,000 cpm each) were added in the 100 μl "marker" aliquot. The precipitate was then counted in a dual channel spectrometer equipped with a punched paper tape printout. A computer program for use on a NCR Century 200 system using intermediate FORTRAN has been written for analysis of both the single and simultaneous radioimmunoassays.

V. CYCLIC NUCLEOTIDE IMMUNOASSAYS

Standard immunoassay curves for cAMP have been reported previously (Steiner et al., 1969, 1970). The antiserum currently used shows a linear displacement of ^{125}I-ScAMP-TME by unlabeled cAMP, plotted as a semilogarithmic function, from 0.25 to 20 pmoles. When the reaction volume is reduced to 150 μl, a fivefold increase in sensitivity is achieved, allowing measurement of 50 femtomoles of cAMP (Fig. 1). Since this degree of sensitivity is rarely necessary, the 600 μl reaction volume is routinely used. Although sensitive cAMP immunoassays can be obtained after only an hour of incubation, we have found the slope of the assay curve to be steeper after approximately 18 hr of incubation, and therefore we usually wait overnight before separating antibody-bound and free cAMP "marker."

Most of the cAMP antibodies exhibit relatively minor cross-reactivity with structurally related nucleotides and nucleosides, and other tissue or plasma constituents. In general, at least a 50,000-fold greater concentration of ATP and other mono-, di-, or triphosphate nucleosides is required to produce displacement of marker equal to that of cAMP. Since tissue ATP levels are generally at least 10,000-fold greater than cAMP levels, it is necessary to select antisera which cross-react < 0.002% with ATP. As seen in Table 1, six of the 10 antisera tested have this degree of specificity against ATP. Cross-reactivity with cGMP in general varies from 1% to 0.01%. Even antibodies with 1% cross-reactivity are suitable for immunoassay since tissue levels of cGMP are

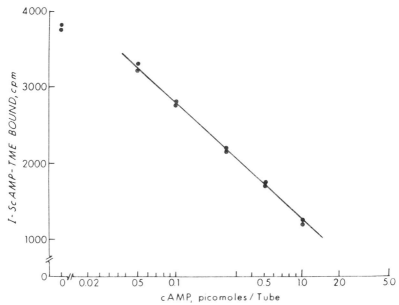

FIG. 1. Standard immunoassay curve for cAMP. Reaction conditions described in text. Reaction volume was 150 μl.

TABLE 1. *Sensitivity and specificity of various cAMP antisera*

Antisera	Maximal sensitivity[a]		Relative binding affinity	
	Serum dilution	cAMP (pmoles/tube)	ATP(%)	cGMP(%)
RcA-1	1 : 5,000	1	0.002	0.01
RcA-2	1 : 5,000	5	0.001	0.005
RcA-3	1 : 5,000	1	0.0001	0.005
RcA-4	1 : 5,000	2	0.001	0.005
RcA-5	1 : 5,000	2	0.0025	0.005
RcA-6	1 : 10,000	1	0.0025	0.005
RcA-7	1 : 5,000	0.25	0.002	0.01
RcA-8	1 : 5,000	5	0.01	0.002
LcA-1[b]	1 : 40,000	0.25	0.0001	1.0
LcA-2[b]	1 : 40,000	0.25	0.0001	1.0

[a]Expressed as the minimal concentration of cAMP which caused linear displacement in the immunoassay procedure described in Methods section.

[b]cAMP antisera obtained from rabbits after three boosts of ScAMP-albumin at monthly intervals kindly supplied by Drs. G. Liddle and N. Kaminsky of Vanderbilt University.

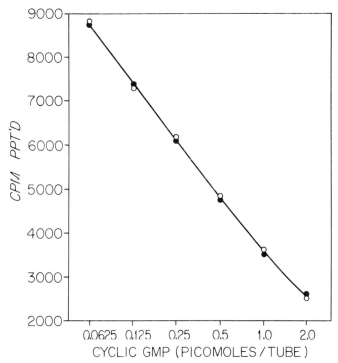

FIG. 2. Standard immunoassay curve for cGMP. o–o represents value obtained by individual radioimmunoassay. ●–● represents value obtained by the simultaneous assay method. Each point represents the mean of three determinations.

in general 2 to 10% of the cAMP level. A standard immunoassay curve for cGMP is shown in Fig. 2. The assay is sensitive to < 0.125 pmoles cGMP. This degree of sensitivity allows measurement of cGMP in triplicate on 20 to 40 mg of most tissues. Cross-reactivity of the cGMP antibody with all purine and pyrimidine nucleotides is minimal (< 0.002%), except for cIMP which reacts at the 1% level.

The sensitivity of the cIMP and cUMP immunoassay is 0.2 and 0.1 pmoles, respectively. Cross-reactivity of the cIMP antibody with all purine and pyrimidine nucleotides was < 0.001% of that with cIMP, except for cGMP and cUMP, which exhibited 1% and 0.14% cross-reactivity, respectively. In the cUMP assay, only cTMP (2%) and cIMP (0.7%) showed any significant cross-reactivity with the antibody.

VI. TISSUE LEVELS OF CYCLIC NUCLEOTIDES

The concentrations of cAMP and cGMP in various tissues from rats and in plasma and urine from humans are shown in Table 2. The concentrations of cAMP and cGMP are in the same range as those reported by others using

TABLE 2. *The concentration of cAMP and cGMP in various tissues and biological fluids*

Tissue	cAMP	cGMP
	(pmoles/g wet wt. of tissue)	
Rat		
liver	960 ± 98 (46) [650 − 1200]	15 ± 2 (44)
kidney cortex	880 ± 92 (61) [720 − 1100]	38 ± 4 (58) [15 − 48]
skeletal muscle	360 ± 52 (22) [120 − 580]	18 ± 1.6 (28) [12 − 26]
isolated fat cells	63 ± 7.4 (78) [30 − 82]	2.4 − 5.3
heart		12 − 24
pituitary[a]	620 − 1400	7 − 8
small intestine mucosa	830 − 970	50 − 160
Human		
lymphocytes[b]	3.1 ± 0.4 (48)	1 − 1.5
plasma[c]	8 − 20	1.8 − 6.0
urine[d]	4.41 ± 0.39 (15) [2.62 − 6.47]	1.09 ± 0.12 (15) [0.6 − 2.6]

Values represent mean ± S.E.M. The number of determinations is given in parentheses and the range of values is recorded in brackets except when less than 10 determinations were carried out—then only the range of values is given.

[a]Hemipituitaries were incubated for 2 hr at 37% in TC 199
[b]Values expressed as pmoles/10^7 cells
[c]Values expressed as pmoles/ml
[d]Values expressed as μmoles/24 hr

enzymatic techniques (Goldberg, Larner, Sasko, and O'Toole, 1969; Goldberg, Dietz, and O'Toole, 1969a; Ishikawa, Ishikawa, Davis, and Sutherland, 1969). In general, the concentration of cAMP in tissues is 20 to 100 times greater than that of cGMP, while in plasma and urine the concentration of cAMP is only three to four times that of cGMP. cUMP and cIMP have not been identified in any rat tissue thus far examined. If present, the concentration of these nucleotides must be less than 1×10^{-8} moles/kg wet weight of tissue.

A convenient means of checking the validity of the cyclic nucleotide determinations is to measure the amount of the immunologically reactive cyclic nucleotide before and after hydrolysis by cyclic nucleotide phosphodiesterase. In the cAMP immunoassay, cyclic nucleotide phosphodiesterase hydrolyzed 85 to 100% of the immunologically reactive cAMP in extracts from various rat tissues and human plasma and urine. In the determination of plasma cAMP, it is important to use an antibody that shows minimal cross-reactivity with uric acid, since antibodies cross-reacting with uric acid give falsely elevated levels of plasma cAMP. This can be checked by performing the appropriate cyclic nucleotide phosphodiesterase control. In the cGMP assay, no "blank" was found in extracts from rat cerebral cortex, pituitary, lung, or intestinal mucosa, or from human urine or plasma; however, a significant "blank" remained in extracts from skeletal muscle (15%), kidney cortex (25%), and liver (40%). In these tissues, the blank is subtracted from the total amount of immunologically reactive cyclic GMP on each sample assayed. Although the exact nature of the material accounting for the cGMP blank has not been determined, it is most likely nucleotide in origin.

VII. SIMULTANEOUS ASSAY FOR cAMP AND cGMP

The standard curve of cGMP obtained in the simultaneous assay is identical to that of the individual cGMP immunoassay (Fig. 2). Identical curves are also obtained for cAMP when performed by simultaneous or individual assay. Moreover, tissue values of cGMP are identical when determined

TABLE 3. *Effect of increasing concentration of cAMP on the measurement of cGMP by simultaneous radioimmunoassay*

cAMP (pmoles/tube)	cGMP (pmoles/tube)	
	Added	Measured
1	0	0.00
10	0	0.00
100	0	0.00
1,000	0	0.44 ± .03
1	0.5	0.53 ± .02
10	0.5	0.52 ± .06
100	0.5	0.51 ± .02
1,000	0.5	0.88 ± .05
1	2.0	1.81 ± .13
10	2.0	1.76 ± .10
100	2.0	2.06 ± .23
1,000	2.0	1.95 ± .16

individually or in the simultaneous assay, as are those of cAMP. Since the tissue concentration of cAMP relative to cGMP is in general at least an order of magnitude greater, it is important to select a cGMP antibody that shows minimal cross-reactivity with cAMP. As shown in Table 3, even a 1,000-fold increase in cAMP relative to cGMP produces minimal changes in the amount of cGMP measured. This permits the determination of both cyclic nucleotides in tissue and body fluids after physiologic stimuli which cause a large change in the concentration of cAMP, for example, without affecting the level of cGMP. When used in conjunction with the computer program, the simultaneous radioimmunoassay technique allows one technician to determine in triplicate both cAMP and cGMP in as many as 300 samples per day.

VIII. DETERMINATION OF ADENYL AND GUANYL CYCLASE ACTIVITY BY RADIOIMMUNOASSAY

As we have reported previously (Steiner et al., 1970), the radioimmunoassay technique has been applied to the measurement of adenyl and guanyl cyclase. In both assays, we measure by radioimmunoassay the amount of cyclic nucleotide generated from tissue fractions incubated with millimolar quantities of the respective nucleotide triphosphate and appropriate cation. In both assays, theophylline is used to inhibit cyclic nucleotide phosphodiesterase. The specificity of the antibodies used in these assays allows for discrimination between the small amount of cyclic nucleotide formed and the relatively high concentration of nucleotide triphosphate required as substrate for each enzyme. Results of adenyl cyclase determination in a variety of tissues are identical when measured by radioimmunoassay or by the isotopic procedure described by Krishna, Weiss, and Brodie (1968). The radioimmunoassay technique is easy to perform, since chromatographic steps are not necessary and expensive radioisotopes are not required.

IX. SUMMARY

We have described in detail the steps necessary for the development of the radioimmunoassay of the cyclic nucleotides. The immunoassays are sensitive to the femtomole range and are easy to perform, since chromatographic steps are not required. The simultaneous radioimmunoassay of cAMP and cGMP employs specific cAMP and cGMP antibodies and both a ^{125}I-labeled cyclic GMP derivative and a ^{131}I-labeled cyclic AMP derivative as markers in the assay tube. Computer programs have been written for both the individual and simultaneous radioimmunoassays. The simultaneous assay permits the determination of both nucleotides in hundreds of tissue and body fluid samples daily.

ACKNOWLEDGMENTS

This work was supported by U.S. Public Health Service grants AM-1921 and A10-2019 from the National Institutes of Health. Dr. Steiner is an Established Investigator of the American Heart Association.

REFERENCES

Falbriard, J. G., Posternak, Th., and Sutherland, E. W. (1967): Preparation of derivatives of adenosine 3',5'-phosphate. *Biochimica et Biophysica Acta*, 148:99.

Goldberg, N. D., Larner, J., Sasko, H., and O'Toole, A. G. (1969): Enzymic analysis of cyclic 3',5'-AMP in mammalian tissues and urine. *Analytical Biochemistry*, 28:523.

Goldberg, N. D., Dietz, S. B., and O'Toole, A. G. (1969a): Cyclic guanosine 3',5'-monophosphate in mammalian tissues and urine. *Journal of Biological Chemistry*, 244:4458.

Greenstein, J. P., and Winitz, M. A. (1961): *Chemistry of the Amino Acids*, Vol. 2. John Wiley and Sons, New York, p. 978.

Hunter, W. M., and Greenwood, F. C. (1962): Preparation of iodine-131 labelled human growth hormone of high specific activity. *Nature*, 194:495.

Ishikawa, E., Ishikawa, S., Davis, J. W., and Sutherland, E. W. (1969): Determination of guanosine 3',5'-monophosphate in tissues and of guanyl cyclase in rat intestine. *Journal of Biological Chemistry*, 244:6371.

Krishna, G., Weiss, B., and Brodie, B. B. (1968): A simple sensitive method for the assay of adenyl cyclase. *Journal of Pharmacology and Experimental Therapeutics*, 163:379.

Steiner, A. L., Kipnis, D. M., Utiger, R., and Parker, C. W. (1969): Radioimmunoassay for the measurement of adenosine 3',5'-cyclic phosphate. *Proceedings of the National Academy of Sciences*, 64:367.

Steiner, A. L., Parker, C. W., and Kipnis, D. M. (1970): The measurement of cyclic nucleotide by radioimmunoassay. In: *Role of Cyclic AMP in Cell Function, Advances in Biochemical Psychopharmacology*, Vol. 3 edited by P. Greengard and E. Costa. Raven Press, New York.

Yalow, R. S. and Berson, S. A. (1960): Immunoassay of endogenous plasma insulin in man. *Journal of Clinical Investigation*, 39:1157.

Advances in Cyclic Nucleotide Research, Vol. 2
Raven Press, New York © 1972

Analysis of Cyclic AMP and Cyclic GMP by Enzymic Cycling Procedures

Nelson D. Goldberg, Ann G. O'Toole, and Mari K. Haddox

*Department of Pharmacology, University of Minnesota,
Minneapolis, Minnesota 55455*

I. INTRODUCTION

It is well recognized that the most troublesome aspect of cyclic nucleotide research centers around the methodology in the field and that the greatest difficulties are those associated with the estimation of endogenous tissue cyclic nucleotide levels. The problems stem from the fact that both cyclic 3',5' adenosine monophosphate (cyclic AMP) and cyclic 3',5' guanosine monophosphate (cyclic GMP) are present in most mammalian tissues in the submicromolar concentration range; and structurally similar, naturally occurring 5' nucleoside phosphates, which represent potential sources of interference in most analytical systems, are present in several hundred to several hundred thousand times the concentration of the cyclic 3',5' nucleotide congenors. A number of sensitive analytical procedures based on a variety of different concepts has developed over the past few years. Advantages and disadvantages in terms of sensitivity, specificity, ease of operation, etc., may be cited for each and investigators today are endowed with a luxury of selecting the method of choice on the basis of their experience and confidence in the particular laboratory skills involved.

Considering the complexity of interrelationships among components in the cyclic nucleotide system that have been elucidated and others that are known to exist but await definition, the availability of different analytical approaches can also be viewed as serving the best interests of the discipline. They should provide a means of uncovering subtleties that might go undetected by a single, universal approach.

In this chapter details of methods based on enzymic cyclic procedures for the estimation of cyclic AMP (Goldberg, Villar-Palasi, Sasko, and Larner,

1967; Goldberg, Larner, Sasko, and O'Toole, 1969) and cyclic GMP (Goldberg, Dietz, and O'Toole, 1969) are presented.

The analytical procedures for both cyclic nucleotides include the following steps:

1. Thin layer chromatographic separation from interfering tissue metabolites.
2. Enzymic conversion of cyclic 3′,5′ nucleotide to 5′ nucleoside di- or triphosphate.
3. Magnification by enzymic cycling.
4. Enzymic-fluorometric detection of enzymic cycling product.

II. SAMPLING AND EXTRACTION

A. Tissues

The tissue levels of cyclic AMP are known to undergo rapid changes seconds after exposure to hormones or as a result of other treatments. Experimental procedures that involve removal of tissue samples from the intact animal can by themselves lead to either rapid decreases or increases in cyclic AMP levels, depending on the conditions and the particular organ or tissue involved. No procedure yet devised has been shown to eliminate such artifacts completely, but at this time there is general agreement that quick-freeze procedures at appropriate times after a given treatment are probably the most acceptable. The use of Wollenberger-type clamps (Wollenberger, Ristau, and Schoffa, 1960)—stainless steel or aluminum blocks cooled in liquid nitrogen which compress and freeze the sample uniformly and efficiently—is preferred. To retain the anatomical integrity of the specimen, the excised sample is rapidly frozen by immersing it in Freon-12 or isopentane cooled to $-150°C$ with liquid nitrogen.

Dissection and weighing, or other preparations of the sample before extraction, should be carried out at temperatures below $-20°C$. The extraction procedure used should minimize losses due to enzymic hydrolysis or continued enzymic or chemical generation of the cyclic nucleotide. For example, phosphodiesterase has been shown to promote cyclic AMP and cyclic GMP hydrolysis at $0°C$ (O'Dea, Haddox, and Goldberg, 1970) at approximately one-third of the rate seen at $30°C$, and divalent cations (e.g., magnesium and barium) are known to promote cyclization of ATP to cyclic AMP at alkaline pH.

Once the activities of the enzymes are permanently arrested, both cyclic nucleotides are relatively stable. The following extraction procedures have been proved to be efficient and reliable.

1. A known weight of frozen tissue sample is powdered and layered over approximately four volumes of frozen 30% methanol containing 10% trichloroacetic acid (TCA), which is conveniently contained in a vessel (glass

microhomogenizer, Micro-Metric Instrument Co., Cleveland, Ohio) that can also be used to carry out the subsequent homogenization and centrifugation. The tubes are then transferred to an ethanol–dry ice bath maintained at -18 to $-17°C$ and the methanol-TCA solution thawed (melting point of this solution, $-18°C$) while the tissue, which remains frozen, settles into the liquified acid solution. After 5 min at $-17°C$, the suspension is briefly homogenized (30 sec), placed on ice for about 10 min, then centrifuged to remove the denatured protein.

2. Four volumes of ice-cold 10% TCA are quickly added to an aliquot of frozen powdered tissue brought to $-20°C$ in an ethanol–dry ice bath, and the sample is rapidly dispersed in the acid by homogenization for about 2 min until the "slush" that results is melted.

^{3}H-cyclic AMP (Schwartz Bioresearch) or ^{3}H-cyclic GMP (Calbiochem), depending on the cyclic nucleotide to be analyzed, is added before TCA is removed from the extracts so that any losses resulting in subsequent steps can be determined. The tritiated cyclic nucleotides should be purified before they are used for this purpose because radioactive impurities sometimes accounting for as much as 15% are present. The purification may be carried out by chromatography on cellulose or silica gel thin-layer plates according to the procedures outlined in Section III. Cyclic AMP should be diluted to approximately 2000 cpm/μl and the tritiated cyclic GMP to about 160 cpm/μl.

Since one-fourth of the total reconstituted volume of the eluate recovered from the thin-layer chromatographic purification of cyclic AMP will be analyzed for recovery (i.e., about 25 μl under the standard assay conditions), 1 μl of the purified ^{3}H-cyclic AMP per 50 μl of acid tissue extract should be added before extraction with ether. This will provide a total of about 500 cpm for counting and a concentration of ^{3}H-cyclic AMP of about 2×10^{-9} M (assuming a specific activity of 16 Ci/mmole). If cyclic GMP is to be determined, 1 μl of ^{3}H-cyclic GMP containing 160 cpm is added per 100 μl of acid extract before the ether extraction procedure. This volume of extract is ordinarily reduced a total of 10-fold (i.e., fivefold concentration before thin-layer chromatography and a twofold concentration of the recovered eluate). This would ultimately provide a total of 200 cpm for counting, and a concentration of about 5×10^{-10} M ^{3}H-cyclic GMP is used in the samples to be analyzed (assuming a specific activity of 24 C/mmole).

Trichloroacetic acid is removed from the clear acid supernatant fraction obtained after centrifugation (10,000 $\times$ g for 20 min) by three successive extractions with 10 volumes of water-saturated ether. Traces of ether are removed by aspirating the vapors above the liquid phase after heating the extract to approximately 60°C in a water bath. Extraction losses of no more than 10% result from ether extraction of the methanol-TCA solution and less than 5% when the 10% TCA solution is used.

The ether-extracted sample (pH 3–4) can be stored at $-80°C$ for an indefinite period of time without loss of cyclic nucleotides.

B. Urine

Because 5′-nucleotides do not normally occur in urine, urine samples may be prepared for analysis by simply heating the sample (90°C for 2 min) immediately after collection, chilling, and clarifying by centrifugation (10,000 × g for 15 min). Either of the cyclic nucleotides in the heat-denatured urine sample may be quantitated without further purification or other treatment of the urine by the analytical methods to be described.

III. THIN-LAYER CHROMATOGRAPHIC SEPARATION OF CYCLIC 3′,5′ NUCLEOTIDES FROM INTERFERING TISSUE COMPONENTS

A. Cyclic AMP

Avicel thin-layer plates (20 × 20 cm, 250 microns thick from Analtech, Inc., Wilmington, Del.) are first washed by development with water, then air dried and activated (90°C for 30 min). Plates are inscribed using a clean microspatula with 15 vertical channels approximately 12-mm wide and with a horizontal line approximately 2.5 cm from the top to halt the migration of the solvent. In addition to the tissue extracts, internal standards of cyclic AMP within them, standards in TCA "blank," and ether-extracted TCA "blank" alone are also chromatographed. The mobility of cyclic AMP is determined by spotting an identical aliquot of tissue extract or ether-extracted TCA "blank" containing approximately 0.01 μmole of cyclic AMP in adjacent channels at one end of the plate, separated from channels containing tissue extracts or internal standards, etc., by an empty channel. Depending on the anticipated cyclic AMP level or the extent to which the original extract volume was reduced, a 10 to 50 μl aliquot of each acid extract is spotted with either a microsyringe or disposable micropipette 3 cm from the bottom of the plate in the center of each channel forming a series of small spots. A clean glass plate raised slightly above the thin layer plate is used to protect the area of cellulose above the portion being spotted. After applying the extract, the thin-layer plates are developed ascending at room temperature in a covered glass tank with approximately 125 ml of freshly prepared solvent composed of redistilled isopropanol, NH_4OH, H_2O in a ratio of 7:1.5:1.5. The solvent is allowed to migrate to the horizontal line 2.5 cm from the top of the plate (approximately 5 hr). The plates are then air dried, and stored in a desiccated chamber until the operation of recovering the cyclic nucleotides is carried out. The R_f values of key tissue intermediates in this chromatographic system are as follows: ATP, ADP, or AMP, < 0.1, and cyclic AMP, 0.47. The R_f values for cyclic AMP chromatographed from tissue extract and ether-extracted TCA "blank" solution are significantly different (0.47 versus 0.51, respectively) so that the ultraviolet detectable cyclic AMP used to monitor its mobility should be chromatographed from both solutions.

B. Cyclic GMP

The procedure for the isolation of cyclic GMP is similar to that described for cyclic AMP, except that silica gel thin–layer plates with fluorescent indicator (20 × 20 cm, 250 microns thick from Analtech, Inc.) are used. The precoated plates are washed in water, dried, and activated as described above. Vertical channels approximately 1.8-cm wide are inscribed to accommodate the large volume usually used, although more concentrated tissue extracts may be chromatographed. Because of the relatively low levels of cyclic GMP, approximately 100 μl of a tissue extract concentrated five-fold, or its equivalent, is usually chromatographed. Ultraviolet-detectable cyclic GMP "markers" in tissue extract and in ether-extracted TCA "blank" solution are also spotted in channels at one side of the plate as described above. The R_f values for GMP, GDP, GTP, cyclic GMP, cyclic AMP, and guanosine are 0.1, 0, 0, 0.52, 0.60 and 0.60, respectively.

The area of cellulose or silica gel containing the tissue or standard samples of the cyclic nucleotide is selected by determining the migration of the authentic substances in the "monitor" channel, which is detectable with a short-wave ultraviolet hand lamp. The segment of each channel (usually not more than 2.5 cm long) corresponding to the authentic cyclic nucleotide is removed by a vacuum device with a Millipore filter (Mitex, No. LSWP-01300) contained in a Swinnex-13 filter holder which serves as a trap for the adsorbent. The adsorbent is eluted with 2 ml of 95% ethanol by connecting the open end of a Luer stub adapter (15 gauge, No. A1030, Clay-Adams, Inc.) attached to the Swinnex holder, to a 5-ml hypodermic syringe (with the aid of a segment of polyethylene tubing) which had been previously filled with 2 ml of 95% ethanol solution. The ethanol solution is allowed to flow slowly (0.5 ml/min) through the adsorbent trapped by the filter and into a 5-ml conical centrifuge tube. About 90% of the cyclic nucleotide is recoverable by this eluting procedure.

The eluates are evaporated to dryness under reduced pressure on a rotary Evapo-Mix (Buchler Inst. Co.) at room temperature. The samples are then reconstituted with a solution appropriate for the analysis of either cyclic AMP or cyclic GMP (see Tables 1 and 2). An aliquot representing approximately one-quarter of the total reconstituted volume is prepared for counting in a scintillation spectrometer with 500 μl of Nuclear–Chicago tissue solubilizer (NCS) and 15 ml of a mixture containing 5% 2,5-diphenyloxazole (PPO) and 0.03% 1,4-bis-2-(4-methyl-5-phenyloxazole) benzene (dimethyl POPOP) in

TABLE 1. *Reagents for the assay of cyclic AMP*

Enzyme step	Reagent[a]	Volume reaction-reagent (μl)	
1. Cyclic AMP → 5'-AMP	Tris-HCl, pH 8, 50; MgSO$_4$, 3; EDTA, 1; reaction initiated with 0.3 μg bovine heart phosphodiesterase[b]	—	25
2. 5'-AMP → ADP → ATP	Tris-HCl, pH 7.7, 225; MgCl$_2$, 10.5; KCl, 150; P-pyruvate, 7.5; dithiothreitol, 15; EDTA, 0.5; ATP, 1 × 10^{-8}M; myokinase,[c] 40 μg/ml; pyruvate kinase,[c] 250 μg/ml	40	15
3. ADP ⇆ ATP	D-glucose, 55; hexokinase,[c] 0.96 mg/ml	45	5
4. Gluc-6-P → 6-P-gluconate	Tris-HCl, pH 7.7; 100; TPN$^+$, 0.1; reaction initiated with 0.5 μg gluc-6-P dehydrogenase	—	1000

[a]Millimolar concentrations in reagent given except where noted.

[b]Reagent 1 was used to dissolve the evaporated cellulose eluate in the case of tissue analysis. When urine (undiluted) is analyzed, 1 to 2 μl of the urine sample is added directly to the reagent.

[c]Protein pellet obtained after centrifugation of the (NH$_4$)$_2$SO$_4$ suspension redissolved in 50 mM Tris-HCl, pH 7.7.

TABLE 2. *Reagents for the assay of cyclic GMP*

Enzyme step	Reagent[a]	Volume reagent-reaction (μl)	
1. Cyclic GMP → 5'-GMP	Tris-HCl, pH 7.7, 50; MgCl$_2$, 2; EDTA, 0.5; reaction initiated with 0.8 μg of phosphodiesterase[b]	11	
2. 5'-GMP → GDP	Tris-HCl, pH 7.7, 222; MgCl$_2$, 12; KCl, 175; ATP, 0.15; creatine phosphate, 1.75; ATP-GMP phosphotransferase, 200 μg/ml; creatine phosphokinase, 0.5 μg/ml	4	15
3. GDP ⇆ GTP	Tris-HCl, pH 7.7, 100; P-pyruvate, 7; MgCl$_2$, 5; sodium succinate, 16; CoA, 2.4; pyruvate phosphokinase,[c] 1.25 μg/ml; succinate thiokinase,[c] 360 μg/ml	2	17
4. Pyruvate → lactate	EDTA, 125; DPNH, 2.2; lactate dehydrogenase, 0.44 μg/ml	4	21
5. DPN$^+$ detection			
Acid destruction	HCl, 5 N	2	23
Strong alkali treatment	NaOH, 8 N	100	123
Detection	H$_2$O	1000	—

[a]Millimolar concentration in reagent given except where noted.

[b]Reagent 1 was used to dissolve the evaporated silica gel eluate in the case of tissue analysis. When urine is analyzed, 1 to 2 μl of the undiluted urine is added directly to the reagent.

[c]Protein pellet obtained after centrifugation of the (NH$_4$)$_2$SO$_4$ suspension redissolved in 50 mM Tris-HCl, pH 7.7.

toluene. A suitable recovery correction is made for each chromatographed sample.

IV. ENZYMIC ANALYSIS OF CYCLIC AMP

The analysis of cyclic AMP is based upon the principles originally set forth by Breckenridge (1964) as modified by Goldberg et al. (1967, 1969). In the procedure, the cyclic AMP isolated by thin-layer chromatography is converted to 5'-AMP with cyclic nucleotide phosphodiesterase (PD) (Reaction 1) and finally to ATP by the combined actions of myokinase (with trace amounts of ATP to initiate the reaction) and pyruvate kinase (PK) (Reaction 2). The ATP generated then serves as the catalytic component in an enzymic cycling system in which equivalent amounts of glucose-6-phosphate (gluc-6-P), proportional to several thousand times the ATP present, are generated as a result of the dismutation between hexokinase and pyruvate kinase (Reaction 3). The gluc-6-P formed is then measured with gluc-6-P dehydrogenase (DH) fluorometrically (Reaction 4).

1)   cyclic AMP $\xrightarrow[\text{H}_2\text{O}]{\text{PD}}$ 5'AMP

2)   5'AMP $\xrightarrow{\text{myokinase}}$ 2ADP $\xrightarrow[\text{P-pyruvate}]{\text{PK}}$ 2ATP
        +   +
     ATP (trace amounts)   2pyruvate

3)

4)   Gluc-6-P $\xrightarrow[\text{DH}]{\text{Gluc-6-P}}$ 6-P-gluconate
        +   +
     TPN$^+$ TPNH + H$^+$

Specificity for cyclic AMP derives from the highly selective action of myokinase with AMP. Under the assay conditions to be described, 5×10^{-14} mole (i.e., 5 μl of a 10^{-8} M solution) of cyclic AMP can be detected reliably with a variation in reproducibility of about 5 %. The procedure described below is designed to accommodate tissue samples of approximately 10 mg which contain about 2.5×10^{-7} mole of cyclic AMP per kg (wet weight). The volumes of the various reaction mixtures in the protocol range from 25 to 50 μl. An increase in "sensitivity" in a practical sense can be achieved by reducing the reaction volumes to concentrate the fixed amount of cyclic AMP, while maintaining the concentration of all other components (e.g., enzymes, co-factors, buffer, etc.). This adjustment will, in effect, increase the Δ sample/Δ blank ratio.

A. Analytical Steps

The composition of the reaction mixture used in each analytical step is shown in Table 1.

Conversion to 5'-AMP. The evaporated samples of cyclic AMP isolated by thin-layer chromatography are reconstituted with 100 μl of the solution designated in Table 1 as the reagent for Step 1. One 25-μl aliquot is removed and prepared for counting in a scintillation spectrometer. Three 25-μl aliquots are transferred to individual glass tubes (7 × 75 mm OD) and 1 μl of beef heart phosphodiesterase (Butcher and Sutherland, 1962 or Boehringer and Sons) diluted appropriately with 25 mM Tris-HCl, pH 8.0, is added to only two of the three samples. After incubation at 37° for 60 min, the reaction is terminated by placing the tubes in a 90° water bath for 4 min.

Comments. In practice the following samples are included: (a) chromatographed tissue extracts, (b) eluates from thin-layer channels spotted with only ether-extracted TCA solution (i.e., "blank"), (c) two or three concentrations (usually 5×10^{-8} M to 1.5×10^{-7} M) of standardized cyclic AMP which has been added to at least two sets of tissue extracts (i.e., internal standards) before thin–layer chromatography, (d) three concentrations of standardized cyclic AMP in the reagent from Step 1, (e) reagent containing no added nucleotide, (f) one or two concentrations of standardized 5'-AMP in the Step 1 reagent, and (g) one or two concentrations of standardized ATP added just before the addition of the cyclic reagent (Step 3) to blank reaction mixtures carried through the entire analytical procedure.

The phosphodiesterase preparation should be analyzed for contaminating adenine nucleotide content before use. Because all preparations of this enzyme are contaminated to varying degrees with 5'-nucleotidase activity, it is important to determine the exact concentration of phosphodiesterase and the time required to complete the hydrolysis of cyclic AMP. Including 5'-AMP standards in the analysis, incubated with and without the phosphodiesterase, will give an approximation of any loss of 5'-AMP (deriving from cyclic AMP) that might occur. If the enzyme concentration and time for the reaction are carefully selected, there will be few or no detectable losses.

Conversion to ATP. The samples are transferred to an ice bath, and 15 μl of the reagent for Step 2, shown in Table 1, is added. The rack of tubes is covered with aluminum foil and the reaction allowed to proceed overnight in a 26°C incubation bath.

Comments. Rabbit muscle myokinase and pyruvate kinase were obtained from either the Sigma Chemical Co. or Boehringer and Sons. Because high salt concentration is very inhibitory to the enzymic conversion and cycling reactions, the $(NH_4)_2SO_4$ concentration in the commercial enzymes is minimized by centrifuging an aliquot of the myokinase and the pyruvate kinase (as well as the hexokinase in the succeeding step) at 12,000 × g for 15 min, removing the $(NH_4)_2SO_4$ supernatant solution with a micropipette, and resuspending the protein pellet with a volume of 50 mM Tris-HCl, pH 7.7 that is equivalent to the volume of $(NH_4)_2SO_4$ solution removed.

All the enzymes used in this procedure (except the gluc-6-P dehydrogenase) and especially the myokinase have been found to be contaminated in varying degrees with adenine nucleotides which can contribute substantially to the reagent blank. This problem has not been as serious lately as in the past, apparently because of more refined enzyme purification procedures employed by commercial supply companies. It is often necessary, however, if a reasonable degree of sensitivity is required, to reduce the concentrations of the contaminating nucleotides. The process is as follows: activated coconut charcoal [50 to 200 mesh (Fisher Chemical Co.)] is treated by five successive sedimentations in 0.1 N HCl, then washed repeatedly with water until a pH of 3 to 4 is achieved. The charcoal is then neutralized by resuspending it in about 10 vol of 25 mM Tris-HCl, pH 7.5, followed by a final wash with water at least three times before it is dried at 90°. To each 100 μl of enzyme solution (resuspended in Tris buffer as described above), 10 mg of the treated charcoal is added and the mixture swirled in a 7 × 75 mm OD tube intermittently for about 15 min at 0°C. The charcoal is then sedimented by centrifugation for 5 min at 5,000 × g. This treatment of the enzymes effectively removes approximately 90% of the contaminating nucleotides, but has little or no effect on the enzyme activity.

Under the assay conditions, the myokinase catalyzed reaction has been found to be complete within 5 hr. However, it is experimentally convenient to allow the reaction to proceed overnight.

The pyruvate kinase concentration in this step is sufficient to satisfy the requirement for the enzymic cycling system in the next step.

Enzymic cycling of ATP. After completing the above reaction, the tubes are thoroughly chilled in an ice bath, and 4 to 5 μl of the D-glucose-hexokinase (yeast, Type C-300, Sigma Chemical Co.) reagent (Step 3) described in Table 1 is added. The reagent should be maintained at 0°C during this procedure. The tubes are then covered with aluminum foil, and transferred to a 37°C water bath. After 60 min, the tubes are placed in a boiling water bath for 2 min and then cooled in an ice bath.

Comments. The $(NH_4)_2SO_4$ suspension of hexokinase should be centrifuged and the protein pellet redissolved in 50 mM Tris-HCl buffer, pH 7.7. Because of the high concentration of enzymes present, it is important to maintain the reaction tubes and reagent at 0°C during the pipetting operation, to insure the same starting time for all tubes.

Addition of standardized ATP to the appropriate blank reactions (in the range of the cyclic AMP standards) at this step provides a means of determining the cycling rate monitoring the completeness of the conversion of 5'-AMP to ATP.

The cycling rate of the system which is linear in respect to time is approximately 3,000 cycles/hr with 5×10^{-8} M ATP at 37°C. The gluc-6-P generated is proportional to the ATP over a wide range of concentrations that includes the lowest concentrations detectable (10^{-9} M) to a concentration that would utilize no more than 90% of the total P-pyruvate present. The kinetic characteristics of this cycling system (and the one used for GDP in the succeeding section) have been described in detail by Cha and Cha (1965, 1970).

Fluorometric measurement of glucose-6-phosphate. To 1 ml of the reagent for Step 4 (Table 1) in a 10×75 mm Pyrex tube, a given aliquot of the cycled reaction from Step 3, predetermined on the basis of the concentration of gluc-6-P expected, is added. There is no need to remove the heat-denatured protein by centrifugation before the gluc-6-P is analyzed fluorometrically. The sensitivity of the fluorometer (Farrand Model A-2 or A-3, primary filters, Corning No. 5860; secondary Nos. 3387 and 4303, Farrand Optical Co.), is set to provide a reasonable ammeter (or galvanometer) deflection within the range of gluc-6-P concentrations to be measured. After an initial reading to determine the blank fluorescence, 1 to 2 μl of gluc-6-P dehydrogenase (yeast, Sigma Chemical Co. or Boehringer and Sons, diluted approximately 1:20 with 0.02% bovine serum albumin) is added. A second ammeter reading to determine the concentration of TPNH generated due to the oxidation of gluc-6-P is taken at the completion of the dehydrogenase reaction. The time course of the reaction should be predetermined with a reaction of known gluc-6-P concentration and with representative samples from the cycled reactions.

Comments. The volume of the cycled reaction analyzed fluorometrically depends on the sensitivity of the instrumental setting and, of course, on the concentration of gluc-6-P present. Ordinarily, about 5 to 10 μl of the cycled reaction is analyzed at a low-to-medium sensitivity setting.

B. Analytical Results

The concentration of gluc-6-P in the aliquot of each sample carried through the entire procedure without phosphodiesterase treatment [i.e., $(-)$ PD] serves as a blank which is subtracted from the average concentration present in the two aliquots exposed to this enzyme. The $(-)$ PD blank values for the different samples are usually identical. The cyclic AMP equivalent for the $(-)$ PD blank with the reagent alone should be below 5×10^{-8} M (based on the level existing at the reaction volume of the first step).

The $(-)$ PD blank, from channels on thin-layer chromatography plates spotted with ether-extracted TCA solution, do not vary appreciably from plate to plate nor are they usually significantly different from comparable blank values obtained with tissue samples. If significant and consistent variations in this value from different types of samples occur (i.e., tissue samples, buffer blank, thin-layer chromatography, etc.) it is an indication of 5'-nucleotide contamination from some component associated with that operation.

A net increment in the gluc-6-P generated as a result of phosphodiesterase addition to buffer blanks [i.e., $(+)$ PD blank] represents possible cyclic AMP contamination of the reagent or contamination of the phosphodiesterase itself with either cyclic AMP or a 5'-adenosine nucleotide.

The values (fluorescence units) obtained for cyclic AMP standards, after subtracting the appropriate blanks, are proportional to the cyclic nucleotide concentration and are conveniently used to construct a standard (linear) curve to determine the concentrations in the tissue extracts. The tissue concentration can be calculated by correcting for recovery, the dilution resulting from the extraction procedure (usually a factor of fivefold) and the thin-layer chromatographic isolation (usually a factor of about twofold). Under the standard assay conditions described here, curves for standards in buffer or tissue extracts are usually almost identical but can on occasion differ in slope by as much as 20 to 25%. Under these circumstances, the internal standard curve should be used as the source of reference.

A typical calculation could be represented as follows:

$$\frac{\Delta(10^{-7} \text{ M})(4.8)(2.2)(1.175)}{100} = \frac{\text{moles} \times 10^{-7} \text{ cyclic AMP/kg tissue}}{\text{(wet weight)}}$$

where 100 represents the number of fluorescent units equivalent to a 10^{-7} M solution of cyclic AMP in the volume of the reaction of the first step; Δ

represents the fluorescence (units) obtained for the unknown tissue extract; 4.8 represents the tissue dilution of 10 mg of frozen powdered tissue with 40 μl of 10% TCA (i.e., assuming that only 80% of the tissue weight is water and only this portion contributes to the total final volume); 2.2 is the correction for the dilution resulting from chromatographing 50 μl of tissue extract and reconstituting the evaporated thin-layer eluate in a total volume of 110 μl; and 1.175 is the correction for the loss of ^{3}H-cyclic AMP (85% recovery).

The results obtained for multiple analysis of a single sample do not differ by more than 10% and usually by less than 5%.

V. ENZYMIC ANALYSIS OF CYCLIC GMP

The procedure described below is similar to the one utilized to provide evidence that cyclic GMP is a naturally occurring constituent of mammalian tissues (Goldberg et al., 1969) and that the metabolic and/or hormonal regulation of cyclic AMP and cyclic GMP and their biological roles (George, Polson, O'Toole, and Goldberg, 1970) may be quite different.

The cyclic GMP isolated by thin-layer chromatography is converted to 5′-GMP with cyclic nucleotide phosphodiesterase (PD) (Reaction 1) and to GDP by the combined actions of ATP-GMP phosphotransferase (GMP kinase) and creatine phosphokinase (CK) (Reaction 2). The amount of GDP generated, is then magnified with the aid of an enzymic cycling system composed of succinate thiokinase (STK) and pyruvate kinase (PK) (Reaction 3) in which pyruvate, equivalent to about 2,000 times the GDP concentration, is generated. The pyruvate is then converted by lactate dehydrogenase (LDH) (Reaction 4) and a stoichiometric amount of DPN$^+$ is produced. After acid destruction of the unreacted DPNH, the DPN$^+$ is converted to a highly fluorescent product by treatment with strong alkali and the fluorescence determined.

The specificity of the system for cyclic GMP stems from the selectivity of GMP kinase for GMP as the nucleoside monophosphate substrate and

1) Cyclic GMP $\xrightarrow[\text{H}_2\text{O}]{\text{PD}}$ 5′GMP

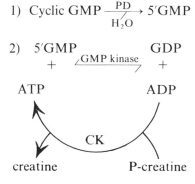

3) P-pyruvate   pyruvate

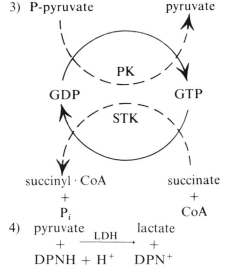

GDP   GTP

succinyl · CoA   succinate
      +   +
      P_i CoA

4)   pyruvate $\xrightarrow{\text{LDH}}$ lactate
        +   +
     DPNH + H$^+$ DPN$^+$

5)   Acid destruction of excess DPNH
6)   Strong alkali treatment of DPN$^+$
7)   Fluorescence measurement of DPN$^+$

succinate thiokinase for guanosine (and inosine) triphosphate. The analytical system provides a level of sensitivity in the range of 10^{-13} to 10^{-14} moles of cyclic GMP which provides for the quantitation of this cyclic nucleotide in 25- to 100-mg samples of most tissue and microliter amounts of urine.

A. Analytical Steps

The composition of the reaction mixture used in each step is shown in Table 2. Analytical details are described below.

Conversion to 5'-GMP. The evaporated eluates obtained from the silica gel thin-layer chromatographic procedure are dissolved in 150 μl of H$_2$O and again evaporated to dryness before they are reconstituted with about 60 μl of the reagent for Step 1 shown in Table 2. Reconstitution and evaporation a second time insures complete recovery from the walls of the tube. One aliquot of about 15 μl is prepared for counting in the scintillation spectrometer. Four 10-μl aliquots are transferred to separate 7 × 75 mm (OD) tubes and 1 μl of purified phosphodiesterase added to only two of the four aliquots, followed by incubation at 37°C for 30 min. The reaction is terminated by heating the tubes in a water bath at 90°C for 4 min.

Comments. The phosphodiesterase prepared from bovine heart by the procedure of Butcher and Sutherland (1962) is more selective for cyclic GMP

($K_{m\ app} = 2 \times 10^{-6}$ M) than for cyclic AMP ($K_{m\ app} = 1 \times 10^{-5}$ M). The V_{max} values with saturating cyclic nucleotide concentrations are similar for both substrates.

The possibility of error due to 5'-nucleotidase in the phosphodiesterase preparation and the extent of the conversion of 5'-GMP to GDP should be evaluated routinely by including 5'-GMP standards.

Three concentrations of cyclic GMP standards in the Step 1 reagent alone (buffer standards) and internal standards which are added to tissue extracts prior to, or after the chromatographic procedure are always included. Eluates obtained from channels on the silica gel which were spotted with only ether-extracted TCA are also analyzed. Extra tubes containing the Step 1 reagent alone should be carried through the procedure so that GTP standards can be added at Step 3 to determine the efficiency of the cycling and conversion systems.

Conversion to GDP. To each tube, 4 µl of the reagent for the second step described in Table 2 is added and the reaction allowed to proceed for 1 hr at 37°C. At the end of the incubation, the rack of tubes is returned to the ice bath for the addition of the next reagent.

Comments. The ATP–GMP phosphotransferase purified by the procedure of Meich and Parks (1965) or obtainable from Boehringer and Sons has never been found to be a source of guanine nucleotide contamination. Although this enzyme is highly specific for GMP, preparations purified by the procedure cited may have activity with 5'-AMP, which is less than 1% of the rate with 5'-GMP. By using creatine phosphokinase (Sigma Chemical Co.) to recycle ADP to ATP in place of pyruvate kinase (Goldberg et al., 1969), the possibility of pyruvate formation from ADP (AMP) is eliminated.

The major source of contamination in this step is guanine nucleotide reactive material present in commercial preparations of ATP to the extent of about 0.05 to 0.5%. Conventional ion-exchange chromatographic purification of the ATP does not remove the interfering substance(s). Cha and Cha (1970) have described an effective procedure for purifying the ATP (and CoA, see Step 3). However, it is also possible to obtain ATP suitable for this procedure by analyzing various commercially available preparations for their guanine nucleotide content (using the analytical procedure described here) and selecting a preparation that is contaminated by no more than 0.05%. Instead of using the excess of ATP shown in Table 2, a concentration of 0.005 mM should be substituted. By carefully selecting the ATP in this manner and reducing the ATP concentration in the reagent, the blank contribution by this component is tolerable (i.e., ca. 2.5×10^{-9} M).

Enzymic cycling of GDP and DPN$^+$ detection. After thoroughly chilling the tubes in an ice bath 2 µl of ice cold cycling reagent, described in Table 2

(Step 3), are added to each tube and the enzymic cycling reaction carried out for 60 min at 37°C. The tubes are then transferred to an ice-water bath, chilled for 10 min, and 4 μl of a solution containing EDTA, lactate dehydrogenase (bovine heart from Worthington) and DPNH added to stop the enzymic cycling reactions by removing the Mg^{++} required by the enzymes in the cycling system and to reduce the pyruvate present to lactate.

The lactate dehydrogenase reaction is allowed to proceed for 15 min at 25°, then 2 μl of 5N HCl are added to each tube to destroy unreacted DPNH. This process is complete within 5 min at room temperature.

The DPN^+ present is then converted to a highly fluorescent product by the addition of 100 μl of 8 N NaOH, followed by an incubation period of 30 min at 38°. An aliquot of each reaction (predetermined by testing representative samples) is transferred to a fluorometer tube (10 × 75 mm) containing 1 ml of glass-distilled water and thoroughly mixed. The fluorescence is then determined in a fluorometer (Farrand, Model A-2 or A-3 with primary filter No. 5860 and secondary filters Nos. 3387, 5563 and 4303).

Comments. Because $(NH_4)_2SO_4$ is very inhibitory to the enzymic cycling reactions, the purified succinate thiokinase and pyruvate kinase, which are both stored as suspensions in 80 to 100% saturated $(NH_4)_2SO_4$, are centrifuged (12,000 × g for 15 min) before use, and the protein pellet remaining after the $(NH_4)_2SO_4$ solution is removed is redissolved in a volume of 50 mM Tris-HCl, pH 7.7, equivalent to the volume of $(NH_4)_2SO_4$ solution removed. Neither of the enzymes has ever been suspected of contamination by guanine nucleotides that might contribute to the blank.

Coenzyme A (CoA), like ATP, is contaminated by "guanine nucleotide-like" compounds but the contaminants can be removed by ion-exchange chromatography. The procedure which was originally described by Cha, Cha, and Parks (1967) is as follows: reduced CoA (approximately 100 μmoles) is oxidized with Lugol's solution (i.e., by titration to a faint orange color) and applied to a DEAE cellulose column in the bicarbonate form (2.5 × 30 cm). After washing with several bed volumes of 0.05 triethylammonium bicarbonate, the oxidized CoA is eluted with a linear gradient (0.05 to 0.5 M) of this salt. The fractions containing the oxidized CoA are pooled and lypholized to remove the triethylammonium bicarbonate which is volatile. Dithiothreitol in 10-fold excess of the oxidized CoA (determined spectrophotometrically at 260 nm, and assuming an A_m of 15.4 × 10^3) is used to reduce the CoA which can then be stored safely at − 80°C for further use.

The equations derived by the Chas (Cha and Cha, 1965, 1970) for the kinetics of enzymic cycling systems are very helpful for establishing guidelines to determine the optimal ratio of activities of the enzymes. The basic equation used (Cha and Cha, 1970) to approximate the cycling rate is as follows:

$$\text{cycles/min} = \frac{v}{S_o} = \left(\frac{K_m + S_o}{V_{max}} + \frac{K'_m + S_o}{V'_{max}}\right)^{-1}$$

where v is the reaction velocity in terms of the rate of formation of pyruvate (μmoles/ml/min); S_o is the sum of the concentrations of the cycling substrate (i.e., GDP and GTP); K_m and K'_m are the apparent values of K_m (μmoles/ml) determined for GDP (ca. 1,200 μM) with pyruvate kinase and for GTP (ca. 10 μM) with succinate thiokinase, respectively; V_{max} and V'_{max} are the apparent maximal velocities (μmoles/ml min) for the pyruvate kinase (with saturating GDP) and succinate thiokinase (with saturating GTP) respectively.

This cycling system can easily be adjusted to provide a 2,000- to 4,000-fold amplification at 37°C in 1 hr. The rate of pyruvate generation is linear with time, until the concentration of CoA becomes rate limiting (i.e., at about 0.05 mM) and proportional to the concentration of GTP (GDP) over a range from below 10^{-9} to 10^{-6} M.

Because most commercial preparations of DPNH are heavily contaminated with DPN^+ (10 to 15%) it is essential that the stock DPNH solution be made up in dilute alkali (50 mM bicarbonate buffer, pH 10.0) and heated before each use to 60°C for about 20 min to destroy the oxidized pyridine nucleotide present. The concentration of DPNH in the stock solution should be no greater than 5 mM. For further details of the stability of pyridine nucleotides the work of Lowry and his co-workers should be consulted (Lowry, Roberts, and Kapphahn, 1957; Lowry, Passonneau, Schulz, and Rock, 1961). The strong alkali treatment of DPN^+ provides about a tenfold increase in the intensity of fluorescence relative to the native fluorescence of a comparable concentration of the reduced form. For this reason, only 15 to 20 μl of the strong alkali-treated reaction is required for the final reading in the fluorometer. In most commercial fluorometers the proportionality between DPNH concentration and fluorescence is linear up to, but not beyond, 2×10^{-5} M DPNH (or its fluorescent equivalent).

B. Analytical Results

The data obtained by this procedure are evaluated in a manner similar to that described for the analysis of cyclic AMP. Ordinarily the reagent blank, $(-)$ PD, is equivalent to a concentration of about 2×10^{-8} M cyclic GMP (determined at the 10 μl volume of Step 1.) Blank $(-)$ PD values do not differ significantly for the different type samples (i.e., tissue extracts, thin-layer chromatography blanks, or buffer blanks). If spurious $(-)$ PD blank values do arise it is, of course, an indication of 5'-guanosine nucleotide contamination. High blank values which are indicative of cyclic GMP contamination are rare.

Because of the relatively large amount of tissue represented by the reconstituted eluates in the analytical procedure, it is not surprising that curves constructed from internal standards differ significantly but consistently (as much as 35 to 40%) from those based on buffer standards. The problem stems from an inhibitory influence on the cycling system by some factor(s) from tissue extracts. It is essential, therefore, that a number of internal standards be included so that correct absolute values may be calculated. In the analysis of urine samples, much less interference is encountered but internal standards represent an important absolute reference.

The variation encountered when a single sample is subjected to repeated analysis is less than 15%, and the variation between duplicate samples in a single experiment is no greater than 5%. The specificity of this system for cyclic GMP has been tested repeatedly. Concentrations of cyclic AMP or 5'-AMP 1000 times greater (i.e., 5×10^{-5} M) are not detectable. Cytidine and uridine nucleotide polyphosphates are also not reactive (Cha and Cha, 1970). Because of the specificity of ATP-GMP phosphotransferase, cyclic IMP and 5'-IMP would not interfere and inosine di- or triphosphate, which could serve as substrates for the cycling system, would be removed from tissue extracts by the chromatographic purification procedure.

REFERENCES

Breckenridge, B. M. (1964): The measurement of cyclic adenylate in tissues. *Proceedings of the National Academy of Sciences*, 52:1580–1586.

Butcher, R. W., and Sutherland, R. W. (1962): Adenosine 3',5'-phosphate in biological materials. I. Purification and properties of cyclic 3',5'-nucleotide phosphodiesterase and use of this enzyme to characterize adenosine 3',5'-phosphate in human urine. *Journal of Biological Chemistry*, 237:1244–1250.

Cha, S., and Cha, C.-J. M. (1965): Kinetics of cyclic enzyme systems. *Molecular Pharmacology*, 1:178–189.

Cha, S., and Cha, C.-J. M. (1970): Microdetermination of guanine ribonucleotides by an enzyme amplification technique. *Analytical Biochemistry*, 33:174–192.

Cha, S., Cha, C.-J. M., and Parks, R. E. Jr. (1967): Succinic thiokinase. IV. Improved method of purification, arsenolysis of guanosine triphosphate, succinate-dependent guanosine triphosphate activity, and some other properties of the enzyme. *Journal of Biological Chemistry*, 242:2577–2581.

George, W. J., Polson, J. B., O'Toole, A. G., and Goldberg, N. D. (1970): Elevation of guanosine 3',5'-cyclic phosphate in rat heart after perfusion with acetylcholine. *Proceedings of the National Academy of Sciences*, 66:398–403.

Goldberg, N. D., Dietz, S. B., and O'Toole, A. G. (1969): Cyclic guanosine 3',5'-monophosphate in mammalian tissues and urine. *Journal of Biological Chemistry*, 244:4458–4466.

Goldberg, N. D., Larner, J., Sasko, H., and O'Toole, A. G. (1969): Enzymic analysis of cyclic 3',5'-AMP in mammalian tissues and urine. *Analytical Biochemistry*, 28:523–544.

Goldberg, N. D., Villar-Palasi, C., Sasko, H., and Larner, J. (1967): Effects of insulin treatment on muscle 3',5'-cyclic adenylate levels *in vivo* and *in vitro*. *Biochimica et Biophysica Acta*, 148:665–672.

Lowry, O. H., Passonneau, J. V., Schulz, D. W., and Rock, M. W. (1961): The measurement of pyridine nucleotides by enzymatic cycling. *Journal of Biological Chemistry*, 236:2746–2755.

Lowry, O. H., Roberts, N. R., and Kapphahn, J. I. (1957): The fluorometric measurement of pyridine nucleotides. *Journal of Biological Chemistry*, 244:1047–1064.

Miech, R. P., and Parks, R. E. Jr. (1965): Adenosine triphosphate: guanosine monophosphate phosphotransferase: Partial purification and substrate specificity. *Journal of Biological Chemistry*, 240:351–357.

O'Dea, R. F., Haddox, M. K., and Goldberg, N. D. (1970): Enzymic hydrolysis of cyclic nucleotides at low temperature. *The Pharmacologist*, 12:291.

Wollenberger, A., Ristau, O., and Schoffa, G. (1960): Eine einfache Technik der extrem schnellen Abkühlung grösserer Gewebestücke. *Pflüger's Archiv für die Gesamte Physiologie des Menschen und der Tiere*, 270:399–412.

Advances in Cyclic Nucleotide Research, Vol. 2
Raven Press, New York © 1972

The Luminescence Assay of Cyclic AMP

Roger A. Johnson*

*Department of Physiology, Vanderbilt University, School of Medicine, Nashville,
Tennessee 37203*

I. INTRODUCTION

For many years, progress in the area of cyclic AMP research has been hampered by the lack of simple, sensitive, analytical procedures for the nucleotide. Recently, however, a wide variety of such procedures has evolved, each with its inherent disadvantages and advantages. The luminescence assay of cyclic AMP was first reported in 1970 (Johnson, Hardman, Broadus, and Sutherland, 1970) and is based on the conversion of cyclic AMP to ATP (Breckenridge, 1964), and the subsequent determination of ATP by its luminescent reaction with firefly luciferin and luciferase (Hastings, 1968). The determination of cyclic AMP by this procedure will be reported here together with some recent modifications.

II. REACTION SEQUENCE

Cyclic AMP is converted to ATP and then determined according to the reaction sequence shown below.

1   cyclic AMP $\xrightarrow[\text{phosphodiesterase}]{\text{3',5'-nucleotide}}$ 5' AMP

2   5' AMP + ATP $\xleftrightarrow{\text{myokinase}}$ 2 ADP

3   ADP + PEP $\xrightarrow{\text{pyruvate kinase}}$ pyruvate + ATP

4   ATP + LH$_2$-luciferase + $\frac{1}{2}$ O$_2$ $\longrightarrow$ L-AMP-luciferase + PP$_i$
                                                + CO$_2$ + H$_2$O + Light

* This work was done during the tenure of a Career Investigator Fellowship of the American Heart Association.

It is apparent from reactions (1)–(3) that AMP or ADP present in the sample or incubation mixture will also be converted to ATP and react with luciferin and luciferase. For this reason, it is necessary that these interfering agents, as well as ATP itself, be removed from the sample prior to assay. Further, to correct for such contamination remaining after purification, each sample is assayed in the presence and absence of phosphodiesterase.

III. SAMPLE PREPARATION

The assay has been used for the determination of cyclic AMP in tissue and urine, and of adenylate cyclase. A purification scheme routinely used is shown below.

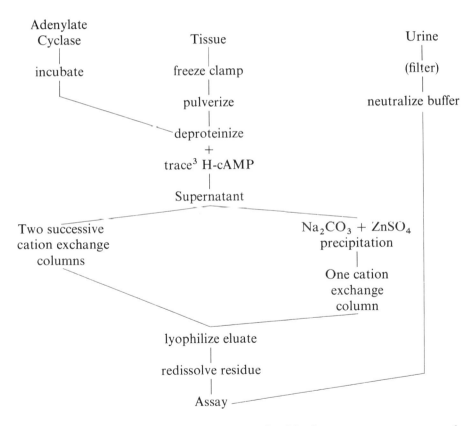

Tissue samples are quick frozen at liquid nitrogen temperature, pulverized, and then deproteinized with 0.3 M perchloric acid containing 5,000 to 10,000 cpm ^{3}H-cyclic AMP to monitor cyclic AMP recovery. Deproteinized samples may be further purified by a number of chromatographic procedures (Goldberg, Larner, Sasko, and O'Toole, 1969; Ishikawa, Ishikawa, Davis, and Sutherland, 1969). One procedure we have used is to apply the sample to a cation exchange column (Bio-Rad AG 50 × 8, 100–200 mesh, H^+-form) and elute with 0.1 N HCl. The fraction containing cyclic AMP is then reapplied, either directly or following lyophilization, to a second cation exchange column. The eluate from the second column is then lyophilized, redissolved, and assayed as described below.

An alternative technique is first to precipitate the noncyclic adenine nucleotides by adding Na_2CO_3 and $ZnSO_4$ (Chan, Black, and Williams, 1970). [For example, to a 1.2 ml sample from an adenylate cyclase incubation are added 100 μl each of 1 M Na_2CO_3 and $ZnSO_4$ (Lin, M.C., *personal communication*)]. The supernatant is then applied to a cation exchange column of sufficient size to remove unprecipitated noncyclic adenine nucleotides and zinc which interferes with the luminescence procedure. The cyclic AMP fraction eluted from the column is lyophilized and the residue redissolved in glycyl-glycine buffer (50 mM, pH 7.5).

Urine samples need only to be filtered and neutralized prior to assay.

IV. ASSAY PROCEDURE

The assay is carried out in two stages in a single disposable culture tube (6 × 50 mm). The conversion of cyclic AMP to ATP involves incubation of the purified sample with the enzymes 3′, 5′-nucleotide phosphodiesterase (Butcher and Sutherland, 1962), myokinase, and pyruvate kinase, all of which are available from Boehringer/Mannheim Corp. Each sample is assayed in duplicate, with and without phosphodiesterase, and with phosphodiesterase plus an added standard (about 10 to 50 pmoles cyclic AMP). This corrects respectively for the presence of contaminating noncyclic adenine nucleotides and for potential assay inhibition by unknown substances. To a 25 μl sample aliquot, with a concentration between 10^{-8} and 10^{-5} M, is added 50 μl of reagent containing: 10 mM $MgSO_4$; 100 mM glycyl-glycine buffer, pH 7.5; 0.1 mM phosphoenolpyruvate (PEP); 10^{-10} M ATP; 2 μg myokinase; 10 μg pyruvate kinase; and, when used, 10 mU phosphodiesterase. The volume is made up to 100 μl with either 25 μl water or external standard, and the resulting solution is capped and incubated overnight at 30°C. The incubated samples can be kept on ice for immediate determination of ATP, or frozen for later analysis.

The determination of the ATP by luminescence is based on reaction 4 which is essentially complete in 1 sec. The peak intensity of light produced is directly proportional to the initial ATP concentration (Fig. 1). Following the initial peak, the light emission decays by a process dependent on numerous

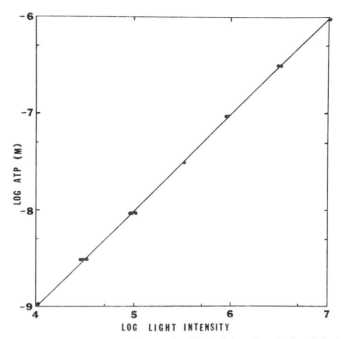

FIG. 1. The luminescence standard curve for ATP. Values for the log light intensity are based on an arbitrary scale.

factors. Because of the nature of the reaction, it is desirable, but not necessary, to measure the initial peak light intensity. Any way in which rapid mixing of the ATP and luciferase can be achieved and the reaction monitored by a sensitive photomultiplier would be satisfactory. The apparatus used may be an instrument designed specifically for luminescence reactions (e.g., Luminescence Biometer from E. I. DuPont de Nemours and Co., Inc., Instrument Products Division), a liquid scintillation counter (Ebadi, Weiss, and Costa, 1970), or a fluorometer with the light source and filters removed. If a fluorometer is used in conjunction with a short rise-time recorder, it is possible to read the peak off the recorder. However, if a liquid scintillation counter is used, it is difficult to record the peak light intensity, and a small area under the curve is usually counted past the peak. Barring interference by substances which change the shape of the light decay curve, the count will be proportional to the initial ATP concentration. Because of the recording problem, it is important that the "counting period" for each sample begin at precisely the same time relative to the mixing of luciferase with ATP. In our hands the most convenient of the alternatives has been to use a Luminescence Biometer since it is designed to store electronically the peak light intensity which can be released later for digital readout. The following description of the procedure for ATP is therefore based on the use of such an instrument.

The incubation tubes in which ATP is generated also serve as cuvettes for the Luminescence Biometer. The cuvette is inserted into the instrument directly in front of the photomultiplier. An aliquot (10 to 20 μl) of luciferin-luciferase reagent is then injected into the tube by means of a 50-μl syringe (Hamilton), to which has been attached a spring-loaded device (Shandon Repro-Jector) which expels the syringe contents at a reproducible velocity. The luciferin-luciferase reagent contains: 100 mM glycyl-glycine buffer, pH 7.5; 10 mM $MgSO_4$; 3 mM dithiothreitol; 1 mg/ml bovine serum albumin; and 20 mg/ml luciferin-luciferase complex. (Crystalline luciferase and synthetically prepared crystalline luciferin have been prepared as a stable enzyme-substrate complex which is available from DuPont.) Inclusion of bovine serum albumin and dithiothreitol is necessary to stabilize the luciferase. The injection velocity is such that adequate mixing results and the peak light intensity can be read off the instrument.

The ATP-luciferin-luciferase luminescence assay is an extremely sensitive and rapid procedure. It is optimally sensitive to about 2×10^{-16} mole ATP and is linear over five orders of magnitude. The range of the assay can be readily adjusted by changing either the luciferase-luciferin concentration or the photomultiplier sensitivity. A typical ATP standard curve commonly used in the assay of cyclic AMP is shown in Fig. 1.

The standard curve for the determination of cyclic AMP by the luminescence assay is shown in Fig. 2. The cyclic AMP standards were incubated in the absence and presence of phosphodiesterase as represented by the two solid lines in the figure. The dashed line represents the difference between these two curves and therefore the ATP generated from the cyclic AMP. The standard curve was linear from 7.2×10^{-9} to 7.2×10^{-6} M cyclic AMP, equivalent to 7.2×10^{-13} to 7.2×10^{-10} moles in the 100 μl sample used in this experiment. The conversion of cyclic AMP to ATP was about 90% complete. The assay blank in this experiment was equivalent to about 2.7 pmoles of cyclic AMP, due principally to contamination by nucleotides bound to the enzymes used to convert cyclic AMP to ATP. The most contaminated of these is myokinase. If one were able to remove the nucleotide contaminants, the sensitivity of the assay would be markedly enhanced. Attempts to remove these nucleotides from the enzymes have employed short charcoal and anion exchange columns (0.48 $\times$ 5 or 10 cm; activated cocoanut charcoal 50–200 mesh; Bio-Rad AG2–X8, 100–200 mesh, Cl$^-$ form) equilibrated and eluted with 5 mM Tris-Cl buffer, pH 7.3. The most effective procedure used both methods and permitted the analysis of 0.15 pmole cyclic AMP.

Dr. Charles A. Sutherland, working in our laboratory, has had more success chromatographing the enzymes on a Sephadex QAE column (A-25, from Pharmacia). The myokinase and pyruvate kinase are first incubated in the presence of phosphoenolpyruvate (PEP) before being applied to a column, previously equilibrated with PEP (cf. reactions (1)–(3)). This converts such

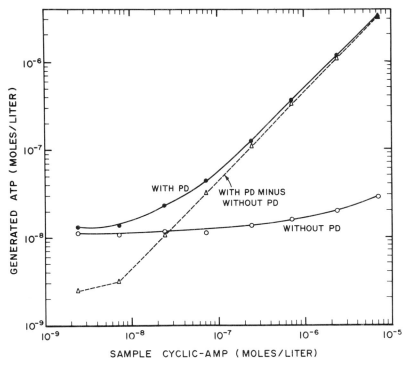

FIG. 2. Standard curve for the assay of cyclic AMP by luminescence. A 200 μl incubation volume was used. Cyclic AMP concentrations are those in the standard solution from which 100 μl aliquots were taken. ATP concentrations are those in the final 200 μl incubation mixture. PD refers to 3', 5'-cyclic nucleotide phosphodiesterase. Myokinase and pyruvate kinase were pretreated with charcoal before use in the incubation mixture. (From Johnson et al., 1970.)

contaminants as AMP and ADP to ATP and thereby facilitates the adsorption of nucleotide to the anion exchanger. In a single passage over this column, the assay blank was reduced by more than 90%, thereby significantly increasing assay sensitivity. The procedure is remarkably simple and should make the assay of samples containing very small amounts of cyclic AMP more practical.

Assay Advantages and Disadvantages

The luminescence assay for cyclic AMP has certain disadvantages. The most important of these are the necessity to remove essentially all noncyclic adenine nucleotides from samples, and, for maximum assay sensitivity, to remove them from the enzymes used in the assay. The second disadvantage is the cost of the instrumentation. Although the assay may be performed using a fluorometer or a scintillation counter, the most convenient instrument to use is the Luminescence Biometer.

The assay exhibits linearity over three orders of magnitude, with a sensitivity to about 1 pmole cyclic AMP. Increasing assay sensitivity tenfold, by removing nucleotide contaminants from the enzymes as described above, would increase the linear range to four orders and the sensitivity to about 0.1 pmole cyclic AMP. The assay specificity is assured by a three–stage discrimination. The first stage of discrimination is the nucleotide precipitation step and cation exchange chromatography. Almost all contaminants would be eliminated by these steps. The second stage is the phosphodiesterase which is specific for the 3', 5'-cyclic phosphate moiety. The third is the luciferase, which is specific for the base moiety of ATP. The procedure is reproducible, generally $\pm 5\%$, and rapid, so that approximately 150 to 200 samples per man-week can be assayed.

Inasmuch as the assay is based on the determination of ATP, a further advantage of the assay is the relative ease with which it could be extended to the measurement of other substances, e.g., nucleotides convertible to ATP. Analogously, the activity of enzymes which either utilize ATP or whose products are convertible to ATP could readily be determined. Such assays for other substances could also be extremely sensitive and rapid and would share many of the advantages described above for the assay of cyclic AMP.

REFERENCES

Breckenridge, B. McL. (1964): The measurement of cyclic adenylate in tissues. *Proceedings of the National Academy of Sciences*, 52 : 1580–1586.

Butcher, R. W., and Sutherland, E. W. (1962): Adenosine 3', 5'-phosphate in biological materials. I. Purification and properties of cyclic 3', 5' nucleotide phosphodiesterase and use of this enzyme to characterize adenosine 3', 5'-phosphate in human urine. *Journal of Biological Chemistry*, 237 : 1244–1250.

Chan, P. S., Black, C. T., and Williams, B. J. (1970): Separation of cyclic 3', 5'-AMP (cAMP) from ATP, ADP, and 5'-AMP by precipitation in adenyl cyclase assay. *Federation Proceedings*, 29 : 616.

Ebadi, M. S., Weiss, B., and Costa, E. (1970): Adenosine 3', 5'-monophosphate in rat pineal gland: Increase induced by light. *Science*, 170 : 188–190.

Goldberg, N. D., Larner, J., Sasko, H., and O'Toole, A. G. (1969): Enzymic analysis of cyclic 3', 5'-AMP in mammalian tissues and urine. *Analytical Biochemistry*, 28 : 523–544.

Hastings, J. W. (1968): Bioluminescence. *Annual Review of Biochemistry*, 37 : 597–630.

Ishikawa, E., Ishikawa, S., Davis, J. W., and Sutherland, E. W. (1969): Determination of guanosine 3', 5'-monophosphate in tissues and of guanyl cyclase in rat intestine. *Journal of Biological Chemistry* 244 : 6371–6376.

Johnson, R. A., Hardman, J. G., Broadus, A. E., and Sutherland, E. W. (1970): Analysis of adenosine 3', 5'-monophosphate with luciferase luminescence. *Analytical Biochemistry*, 35 : 91–97.

Advances in Cyclic Nucleotide Research, Vol. 2
Raven Press, New York © 1972

Firefly Luminescence in the Assay of Cyclic AMP

Manuchair S. Ebadi

*Department of Pharmacology, College of Medicine, University of Nebraska,
Omaha, Nebraska 68105*

I. INTRODUCTION

Bioluminescence and chemoluminescence have been used for determination of picomole levels of adenine dinucleotide, flavin mononucleotide (Stanley, 1971), adenosine diphosphate, and adenosine triphosphate (McElroy and Strehler, 1949; Holmsen, Holmsen, and Bernhardsen, 1966; St. John, 1970). In addition, bioluminescence has been used to determine the activity of a variety of enzymes such as phosphodiesterase (Weiss, 1971; Ebadi, in press), ATP sulphurylase (Balharry and Nicholas, 1971), pyridoxal phosphokinase (Ebadi, in press), and many ATP-requiring enzymes such as myokinase, pyruvate kinase, creatine kinase, nucleotide phosphokinase, and hexokinase (see Strehler, 1968 for review). This chapter discusses the use of the firefly luciferin–luciferase system for the specific and sensitive determinations of the subpicomole levels of adenosine 3′,5′-monophosphate (cyclic AMP) in urine and various tissues (Ebadi, Weiss, and Costa, 1971a).

II. PRINCIPLE

This method is based on the isolation and purification of cyclic AMP, followed by its conversion to ATP by a combined system of phosphodiesterase, myokinase, and pyruvate kinase. The concentration of the resulting ATP is determined in a liquid scintillation counter utilizing the luciferin–luciferase system.

III. METHODOLOGY

A. Preparation of Stock Solutions of Various Nucleotides

Adenosine triphosphate (ATP), adenosine diphosphate (ADP), adenosine 3',5'-monophosphate (cyclic 3',5'-AMP), and adenosine 5'-monophosphate (5'-AMP) were purchased from Sigma Chemical Company (St. Louis, Mo.). Cyclic [^{3}H] AMP (specific radioactivity, 2.35 C/mmole) was obtained from Schwartz Bioresearch, Inc. (New York, N.Y.). Stock solutions of ATP, ADP, 5'-AMP and cyclic 3',5'-AMP with the concentration of 1×10^{-3} M were prepared in Tris-HCl (pH 7.5) and frozen at $-20°$C. As needed, working standards (1×10^{-6} M) were prepared by diluting the stock solution in the same buffer. Cyclic [^{3}H], 3',5'-AMP, which is used to monitor the recovery of cyclic AMP (see parts F and H, under Methodology), was dissolved in 0.3 M ZnSO$_4$.

B. Preparation of Cation Exchange Resin

Analytical grade cation exchange resin AG $50 \times$ W-X8 (200–400 Mesh, H$^+$ form) was purchased from Bio-Rad Laboratories (Richmond, Calif.). A quantity of the resin, usually 1 lb, was washed in 1 liter of 1 N HCl. After the acid was removed, the resin was washed repeatedly with distilled water until the pH of the resin and of the distilled water became equal. The resin was then kept in a refrigerator under water and adjusted appropriately so that the volume of sedimented resin would be exactly one-third of the total volume. Immediately before use, the resin was dispersed by a thorough mixing. A desired portion was transferred to a beaker and continuously mixed by a magnetic stirring apparatus. This technique not only facilitates the task of packing the columns uniformly, but also greatly adds to the accuracy and the duplicability of the results (See part F, under Methodology).

C. Purification of Phosphodiesterase

The enzyme 3',5'-cyclic nucleotide phosphodiesterase which is usually isolated from beef heart may be purchased from Sigma Chemical Company (St. Louis, Mo.). One unit of this enzyme converts 1.0 μmole of 3',5'-cyclic AMP to 5'-AMP per minute at pH 7.5 at 30°C. Other enzymes associated with this phosphodiesterase are 5'-nucleotidase, 5'-ATPase, inorganic pyrophosphatase, and alkaline phosphatase. Most investigators convert cyclic AMP to 5'-AMP by phosphodiesterase, terminate the reaction after a few minutes, and convert the resulting 5'-AMP to ATP. In this case the associated contaminating enzymes present no problem. We have chosen to convert cyclic AMP to 5'-AMP, and the 5'-AMP in turn to ATP in a single incubation period of 4 to 6 hr. Consequently, the accompanying 5'-nucleotidase and 5'-ATPase present with Sigma's preparation will result in lower ATP production. Furthermore, the pyrophosphatase will interfere with the accuracy of luciferase assay (see part H, under Methodology).

A fairly pure phosphodiesterase is essential. This enzyme was purified from hog cerebral cortex according to Nair (1966). Nair's method is based on the classical technique of enzyme purification which involves ammonium sulfate and ethanol fractionations of the homogenate followed by DEAE cellulose column chromatography. This enzyme was shown to have consistent purity several times in our laboratory, although with varied specific activity. The tissue level of cyclic AMP has been determined by using cyclic nucleotide phosphodiesterase with a specific activity of 23 mmoles of cyclic AMP

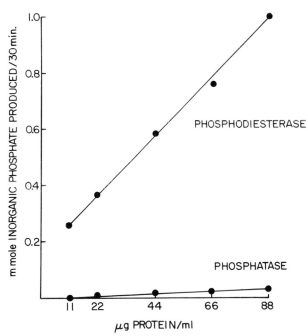

FIG. 1. The relative activity of a nonspecific alkaline phosphatase present as a contaminant in the purified preparation of pig brain cyclic 3',5'-nucleotide phosphodiesterase. Phosphodiesterase was purified from pig cerebral cortex according to the method of Nair (1966) and its activity was determined by the method of Butcher and Sutherland (1962). In assaying cyclic AMP, concentrations of protein smaller than 22 $\mu g/\mu l$ were used. At this concentration, the relatively negligible amount of 5'-nucleotidase, 5'-ATPase, inorganic pyrophosphatase (not shown), and alkaline phosphatase will not interfere with the formation or stability of ATP during the 6 hour period of incubation. (Reprinted with permission from Ebadi et al., 1971a.)

hydrolyzed/mg of protein per 30 min at 30°C. The preparation was stored in concentrated form (45 mg protein/ml) at −45°C. As needed, a portion was diluted 1:20 (v/v) with 50 mM Tris-HCl (pH 7.5) containing 3 mM MgSO$_4$. This stock solution was then divided into 1-ml portions, kept frozen at −20°, and used as needed. Under our assay condition (see part G, under Methodology), phosphodiesterase in the concentration of 11.0 μg of protein/ml could convert 0.25 mmole of cyclic AMP to 5'-AMP in 30 min at pH 8.0. The relative activity of alkaline phosphatase was present in negligible amounts as a contaminant in the purified phosphodiesterase (Fig. 1). Consequently, the enzyme purified in our laboratory is purer and more active than that supplied by Sigma.

It is highly recommended that the investigators who wish to use Sigma's phosphodiesterase and convert cyclic AMP to 5'-AMP and ATP in a single reaction as described in this chapter carefully assess the optimum amount of the enzyme needed for maximum conversion of cyclic AMP to 5'-AMP and minimize the hydrolysis of the latter compound to adenosine and inorganic phosphate.

D. Decontamination of Myokinase and Pyruvate Kinase

Myokinase (ATP:AMP phosphotransferase; EC 2.7.4.3) and pyruvate kinase (ATP:pyruvate phosphotransferase; EC 2.7.1.40) were purchased from Boehringer Mannheim and Sons (New York, N.Y.). These enzymes, which are suspended in (NH$_4$)$_2$SO$_4$, are heavily contaminated with interfering nucleotides, especially ATP which requires decontamination by charcoal treatment according to the method of Goldberg, Larner, Sasko, and O'Toole (1969). Activated coconut charcoal (50–200 mesh, Fisher Chemical Co.) was suspended and washed several times in five volumes (w/v) of 0.1 N HCl. The HCl-treated charcoal was washed with water repeatedly until it assumed the pH of the distilled water. It was then neutralized with 20 mM Tris-HCl (pH 7.5) and washed with distilled water, after which it was dried in an oven at 90°C. A portion was removed and brought to a temperature of 0 to 3°C prior to its use.

The desired quantities of myokinase and pyruvate kinase were transferred to a test tube and centrifuged at 12,000 × g for 15 min. After gently decanting the supernatant fluid, the packed enzyme was resuspended in 50 mM of Tris-HCl (pH 7.5), equivalent to the volume of (NH$_4$)$_2$SO$_4$ removed. Cold charcoal (200 mg charcoal/cc enzyme) was slowly added to the enzyme and gently mixed. Immediately after the charcoal is added, a white effervescent bubbling results. The mixture was kept in crushed ice and gently mixed at required intervals (two or three times) for 15 min. The enzyme was then centrifuged for 10 min at 5,000 × g. The charcoal solutions containing myokinase and pyruvate kinase were aspirated into their original washed container. This decontamination process, although it removes the majority of the bound nucleotides, will also result in a significant loss of activity.

Treatment of myokinase and pyruvate kinase with Dowex-2 has been reported to be superior to charcoal treatment, in that assay blanks are drastically reduced with no apparent loss of activity to the enzymes (Johnson, Hardman, Broadus, and Sutherland, 1970).

E. Preparation of "Conversion" Buffer

The isolated cyclic AMP (see part F, under Methodology) is converted to ATP in a reaction mixture called "conversion" buffer. This buffer, which is a modification of the one prepared by Breckenridge (1965), contains 100mM Tris-HCl (pH 7.5), 50 mM KCl, 2 mM $MgCl_2$, 0.1 mM EDTA, 15 mM dithiothreitol, 0.01% bovine serum albumin, 0.26 mM phosphoenol-pyruvate, 1×10^{-12} M ATP, phosphodiesterase, pyruvate kinase, and myokinase. All chemicals must be in the purest form available. The concentrations of the enzymes is discussed in part G of this section. It is recommended that a buffer containing 100 mM Tris-HCl first be prepared and adjusted to pH 7.5. Then, KCl, $MgCl_2$, EDTA, dithiothreitol, albumin, and phosphoenol pyruvate are weighed separately and added. These will not alter the pH of the buffer. A stock solution of ATP must be made and serially diluted to finally contain 1×10^{-9} M of ATP. One cc of this solution is added to the buffer which is adjusted to a volume of 1 liter, thoroughly mixed, divided into small quantities (e.g., 20 cc), and kept frozen. As needed, a portion is thawed and used. The myokinase, pyruvate kinase, and phosphodiesterase are added separately (part G under Methodology).

F. Preparation of Tissue Extracts and Isolation and Purification of Cyclic AMP

In any particular tissue, the concentration of cyclic AMP depends, among many factors, on the relative activity of adenyl cyclase (the synthesizing enzyme) and phosphodiesterase (the catabolizing enzyme). Since the variable lability and activity of these enzymes results in rapid postmortem changes in the concentration of cyclic AMP (Ebadi, Weiss, and Costa, 1971a,b), the tissues must be processed immediately. In broken cell preparations, $ZnSO_4$ inhibits the activity of phosphodiesterase (Yamamoto and Massey, 1969) and adenyl cyclase (Sutherland, Rall, and Menon, 1962). Therefore, we took advantage of this property and homogenized the frozen tissue in $ZnSO_4$ solution.

The animals (male, Sprague-Dawley rats, weighing 180 to 220 g) were decapitated, and the desired tissues were frozen immediately on dry ice or in liquid nitrogen and kept frozen until assayed. A small portion of the desired tissue, usually 2 to 5 mg, is homogenized in a Duall Glass tissue grinder with

300 μl of 0.3 M $ZnSO_4$ containing 1×10^{-13} moles of $[^3H]$ cyclic AMP (2.35 Ci/mmole) to monitor recovery. A Duall Glass tissue grinder (Kontes Glass Company, Vineland, N.J.) is most useful when homogenizing small amounts of soft tissues such as the brain. However, large and hard tissues such as the entire heart should be frozen in liquid nitrogen and powderized before homogenizing. Mincing the hard tissue will take too long and drastically alters the level of cyclic AMP. The cyclic AMP is mostly particulate-bound (Ebadi et al., 1971b) and must be liberated. This may be accomplished by repeated freezing-thawing (usually by placing the homogenizing tubes in acetone containing dry ice) and rehomogenizing between each freezing and thawing process. The homogenized sample contains many interfering substances (coenzyme A, NAD, $NADH_2$, NADP, $NADPH_2$, AMP, ADP, and ATP) which must be removed. The decontamination process is carried out by $Ba(OH)_2$ precipitation and column chromatography as described next.

After repeated freezing-thawing, a small portion of the homogenate, usually 20 μl, was stored for determination of protein by the method of Lowry, Rosebrough, Farr, and Randall (1951). Nucleotides other than cyclic AMP were precipitated by addition of 280 μl of 0.25 M $Ba(OH)_2$. After centrifugation (International centrifuge, at 6,000 rpm for 5 min), 50 μl each of $ZnSO_4$ and $Ba(OH)_2$ were added to the supernatant, mixed, and recentrifuged as described above. These purification steps should be carried out as rapidly as possible, since cyclic AMP can be formed nonenzymatically from ATP in aqueous $Ba(OH)_2$ digests. Thus Cook, Lipkin, and Markham (1957) have shown that boiling ATP in 0.4 N $Ba(OH)_2$ for 30 min resulted in the conversion of 5 to 10% of the ATP into cyclic AMP. We found that if the samples were exposed to $Ba(OH)_2$ for 1 hr at 25 to 30°C, there was about 0.1% conversion of ATP to cyclic AMP. However, if the samples were processed rapidly and kept cool until they were placed on the cation exchange column and if a slight excess of $ZnSO_4$ was added (to minimize free barium ions), less than 0.025% conversion of ATP to cyclic AMP occurred. Our method of assaying for cyclic AMP automatically corrects for this small conversion (see part H under Methodology). This problem may be totally avoided by following the technique of Chan, Black, and Williams (1970) who have shown that 5'-AMP, ADP, and ATP, but not cyclic AMP, are precipitated by a combination of $ZnSO_4$ and $BaCl_2$, Na_2CO_3 and $BaCl_2$, Na_2SO_4 and $BaCl_2$, Na_2CO_3 and $CdCl_2$ or Na_2CO_3 and $CaCl_2$. Recovery of cyclic AMP by using any of these combinations was greater than 97%. With $ZnSO_4$-$Ba(OH)_2$, however, they obtained 81% recovery.

The supernatant fluid obtained after the second $ZnSO_4$-$Ba(OH)_2$ treatment is still contaminated with adenosine and trace amounts of 5'-AMP, ADP, and ATP, and therefore, requires further chromatographic purification. Glass columns (inside diameter, 0.5 cm, Kontes Glass Co., Vineland, N.J.) were packed with AG50 W-X8 cation exchange resin (200–400 mesh, H^+ form) to a depth of 3.0 cm. This was accomplished by pouring 3 cc of resin slurry (part B, under Methodology) directly into the column. After washing the columns with 10 ml of distilled water, 0.5 ml samples of the supernatant fluids were placed on the columns, which were then eluted with several 2 ml portions of distilled water. ADP and ATP were eluted upon addition of the first 2 ml and were discarded. Cyclic AMP was eluted in the next fraction after another 3 ml portion was added and collected in scintillation counting vials. AMP could be eluted between fractions 6 to 12 ml, whereas adenosine remained on the columns (Krishna, Weiss, and Brodie, 1968). Since there is considerable variation in the properties of resins obtained from different sources and even for different batches secured from a single supplier, the differential elution procedure must be monitored carefully, either by chromatographing a mixture of the nucleotides already listed and determining their absorption at 260 nM or by chromatographing a mixture of labeled nucleotides and determining their respective radioactivities. We stress this recommendation, since early elution may yield interfering nucleotides, whereas late elution may result in poor recovery of cyclic AMP.

Standards of cyclic AMP (10 to 200 pmoles) were homogenized in $ZnSO_4$ in the presence and absence of tissues and were carried through the entire procedure as described (Table 1).

TABLE 1. *Determination and recovery of cyclic AMP in rat cerebellum in the presence and in the absence of internal standard*

Substances assayed	Test (cpm)	Blank (cpm)	pmoles of cyclic AMP	Recovery of internal standard
blank		240		
10 pmoles of cyclic AMP	3,200		10	
20 pmoles of cyclic AMP	8,820		20	
50 pmoles of cyclic AMP	28,800		50	
100 pmoles of cyclic AMP	76,450		100	
5 mg of cerebellum	8,800	1,800	18	
5 mg of cerebellum + 10 pmoles of cyclic AMP	12,600	1,850	25	70%
5 mg of cerebellum + 50 pmoles of cyclic AMP	42,000	1,750	65	94%
10 mg of cerebellum + 50 pmoles of cyclic AMP	61,080	2,400	80	88%

Various concentrations of cyclic AMP (10–50 pmoles) were homogenized in the presence and absence of cerebellar tissue. Cyclic AMP was isolated and converted to ATP in a reaction mixture containing 500 μl of conversion buffer and 5 μl of an enzyme mixture containing 2.5 μl of phosphodiesterase, 15 μl of myokinase, and 40 μg of pyruvate kinase. The reaction was carried out for 6 hr and terminated by addition of 20 μl of 6% (v/v) H_2O_2. The resulting ATP was determined using 400 μg of firefly luciferin-luciferase (part H, Methodology). (Reprinted with permission from Ebadi et al., 1971a.)

G. Conversion of Cyclic AMP to ATP

The cyclic AMP (fraction 2 to 5 ml) was collected in scintillation counting vials, divided into two equal portions (blank and test), and freeze-dried. To each vial, 500 μl of "conversion" buffer (see part E, under Methodology) was added, and the vials were gently rotated to dissolve any cyclic AMP adhering to the walls of the containers. The test reactions were started by addition of 5 μl of a mixture containing 2.5 μg of phosphodiesterase, 40 μg of pyruvate kinase, and 15 μg of myokinase. The tissue blanks received a mixture containing myokinase, pyruvate kinase, and either boiled or no phosphodiesterase. The vials were mixed and placed in a rubber test tube rack with compartment diameter of 3.2 × 3.2 cm (Will Scientific, Inc., Columbus, Ohio). Because 500 μl is not sufficient volume to cover completely the bottom of the scintillation counting vials, the reaction mixture will not be uniformly mixed. To avoid this problem, the vials were placed at an angle of 45 degrees and held in this

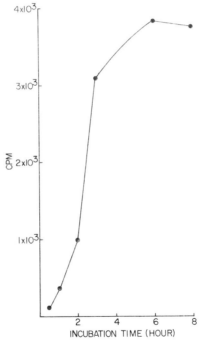

FIG. 2. The relationship between incubation time and production of ATP. The reaction mixtures contained 40 pmoles of cyclic AMP, 500 μl of conversion buffer, and 5 μl of a mixture containing 2.5 μg of phosphodiesterase, 15 μg of myokinase, and 40 μg of pyruvate kinase. The reactions were carried from 15 min to 8 hr and were terminated by addition of 20 μl of 6% (v/v) H_2O_2. The samples were frozen, thawed, and assayed for ATP (part H under Methodology). (Reprinted with permission from Ebadi et al., 1971a.)

position by paper tape. The reaction mixtures were incubated at 30°C in a Dubnoff metabolic shaking incubator (Precision Scientific Co., Chicago, Ill.) for 6 hr which converted all of cyclic AMP to ATP (Fig. 2) according to the following reaction:

$$\text{Cyclic 3',5'-AMP} + H_2O \xrightarrow[\text{phosphodiesterase}]{\text{Cyclic 3',5'-nucleotide}} \text{5'-AMP} + H^+$$

$$\text{5'-AMP} + \text{ATP} \xrightleftharpoons{\text{myokinase}} \text{2ADP}$$

$$\text{ADP} + \text{phosphoenol pyruvate} \xrightarrow[Mg^{++}]{\text{pyruvate kinase}} \text{ATP} + \text{pyruvate}$$

The reactions were terminated by addition of 20 μl of 6% (v/v) of H_2O_2 and the vials were shaken and frozen.

The amounts of phosphodiesterase, myokinase, and pyruvate kinase

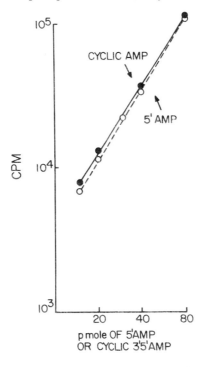

FIG. 3. The rate of production of ATP from 5'-AMP or from cyclic AMP. The reaction mixtures contained 5–80 pmoles of either 5'-AMP (○) or cyclic AMP (●), 500 μl of conversion buffer, and 5 μl of a mixture containing 25 μg of phosphodiesterase, 15 μg of myokinase, and 40 μg of pyruvate kinase. The reactions were carried out for 6 hr and were terminated by addition of 20 μl of 6% (v/v) H_2O_2. The samples were frozen, thawed, and assayed using 500 μg firefly luciferin–luciferase (part H under Methodology). (Reprinted with permission from Ebadi et al., 1971a.)

required were determined empirically to assure complete conversion of all the cyclic AMP to ATP during the 6-hr period of incubation (Goldberg et al., 1969).

It is strongly recommended to determine the extent of contamination of phosphodiesterase (with 5′-nucleotidase, 5′-ATPase, inorganic pyrophosphatase, and nonspecific alkaline phosphatase) and the amount of cyclic AMP lost through $ZnSO_4$–$Ba(OH)_2$ precipitation and column fractionation by converting 10 to 200 pmoles directly to ATP without the above-mentioned treatment. The activity of the charcoal-treated or Dowex-treated myokinase and pyruvate kinase should be checked using a standard concentration of

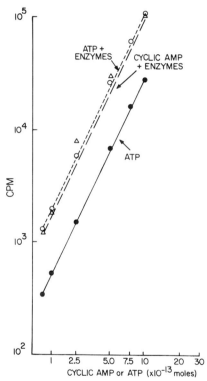

FIG. 4. The production of ATP from cyclic AMP. The reaction mixtures contained 5×10^{-14} to 10^{-12} mole of cyclic AMP ($\Delta \cdots \Delta$), 500 μl of buffer, and 5 μl of an enzyme mixture containing 2.5 μg of phosphodiesterase, 15 μg of myokinase, and 40 μg of pyruvate kinase. The reactions were carried out for 6 hr and were terminated with 20 μl of 6% (v/v) H_2O_2. The reaction mixtures were transferred to a culture tube (0.5 × 5 cm) and lyophilized. The ATP was taken up in 90 μl of 50 mM Tris-HCl (pH 7.5) containing 3 mM $MgSO_4$ and assayed using 100 μg of firefly luciferin–luciferase in a final vol of 100 μl (part A under Discussion). Light emission was also determined for standard concentration of ATP (5×10^{-14} to 10^{-12} mole) in a reaction mixture of 100 μl in the presence ($O \cdots O$) and in the absence ($\bullet \cdots \bullet$) of 5 μl of the enzyme a mixture containing 2.5 μg of phosphodiesterase, 15 μg of myokinase, and 40 μg of pyruvate kinase. (Reprinted with permission from Ebadi et al., 1971a.)

5'-AMP (10 to 200 pmoles) converted to ATP as described. To examine the stability of ATP in the reaction mixture during incubation, various concentrations of this nucleotide (10 to 200 pmoles) were incubated in the presence and absence of phosphodiesterase, myokinase, and pyruvate kinase. Incubation of cyclic AMP with the enzyme system gave the same ATP value as incubation with 5'-AMP (Fig. 3), an indication of complete conversion of cyclic AMP to 5'-AMP. Cyclic AMP yielded the same value as ATP (Fig. 4), indicating the complete conversion of cyclic AMP to ATP.

H. Determination of ATP in a Liquid Scintillation Counter

Light emission generated by the firefly luciferin-luciferase system has been used for the specific and sensitive determination of ATP (Seliger and McElroy, 1965; McElroy, 1947; Strehler and Totter, 1952; Addanki, Sotos, and Rearick, 1966; Lyman and DeVincenzo, 1967) according to the following reactions:

$$\text{luciferase} + \text{D-luciferin} + \text{Mg}^{2+} + \text{ATP} \rightleftharpoons \text{luciferase-luciferyl adenylate}$$
$$+ \text{Mg}^{2+} + \text{PPi}$$

$$\text{luciferase-luciferyl adenylate} + \text{O}_2 \rightarrow \text{luciferase} + \text{CO}_2 + \text{AMP}$$
$$+ \text{thiazolinone} + \text{light}$$

The light produced may be detected by a number of photoelectric devices such as the apparatus of Anderson, Farrand photoelectric fluorimeter, Macnichol's photomultiplier photometer, recording light integrator utilizing vacuum tube electrometer, Turner fluorometer model 110, Aminco photometer-photomultiplier, and Aminco Bowman spectrophotometer (Chase, 1960; Strehler, 1968), by a luminescence biometer (Johnson et al., 1970), or by any liquid scintillation spectrometer (Ebadi et al., 1971a).

The frozen reaction mixtures containing the ATP (part G under Methodology) are thawed and brought to 25°C. To each tube, 500s μl of 50 mM Tris-HCl buffer (pH 7.5) containing 3 mM MgSO_4 was added and the contents of the vials were thoroughly mixed. At zero time, 300 to 600 μg of firefly lantern extract (purchased from Sigma) and reconstituted according to the manufacturer's recommendation) was added, mixed, and placed rapidly (within 5 sec) into the counting well of any scintillation counter which was preset to produce maximum amplifier gain (Stanley and Williams, 1969). At 20 sec after the addition of luciferin-luciferase extract, the samples were counted for a period of 30 sec. The number of counts accumulated in that time is the criterion employed to measure light emission produced by utilization of ATP (Tal, Dikstein, and Sulman, 1964). The counts from the standard samples of cyclic AMP were plotted on full logarithmic paper and the concentration of ATP in the test and blank were individually determined from the curve. Then 15 ml of liquid scintillation fluid (Bray, 1960) was added, and the samples were recounted to calculate the recovery of cyclic [^{3}H] AMP in each sample (part F under Methodology).

Factors influencing the luciferin-luciferase reactions. In addition to luciferin, luciferase, ATP, and a bivalent ion such as magnesium or manganese (McElroy and Strehler, 1949), oxygen is necessary for the light emitting reactions (McElroy and Seliger, 1961). The firefly enzyme is inhibited by a monovalent ion such as chloride, phosphate buffers, high magnesium concentration at low ATP, and even ATP itself at low magnesium concentration (St. John, 1970). Maximum luminescence is observed at an oxygen concentration between 0.1 and 0.5%. This requirement may be clearly checked by deaerating and degassing the reaction mixture prior to initiation of reaction.

The optimum pH of the reaction at 25°C is 7.4 (McElroy and Seliger, 1961). This exact requirement is of crucial importance, since by decreasing the pH from 7.4 to 7.0, 6.5, and 6.0, the intensity of luminescence is decreased from 100% to 52, 12, and 0.1%, respectively. Elevating the pH is also detrimental, since the enzyme maintains only 28% of its original activity at pH 8.0.

Similarly, the optimum temperature for the reaction is 25°C. By decreasing the temperature from 25° to 20°, 15°, and 3°C, the intensity of luminescence is decreased from 100% to 90, 60, and 20%, respectively. Elevating the temperature produces similar qualitative reduction in luminescence, since the enzyme maintains less than 10% of its activity at a temperature of 45°C.

Freshly suspended crude firefly extracts, which contain ATP and ATP-regenerating systems, emit logarithmically increasing amounts of light (Strehler, 1968). Therefore, not only must the blanks be accounted for, but the concentration of firefly extract must be kept constant at a minimum (300 μg in 1 ml of reaction mixture). The luciferin–luciferase provided by Worthington (Freehold, N.J.) seems to be homogenous after its reconstitution with water and is extremely active, but yields a high blank. The luciferin-luciferase provided by Sigma produces a suspension after its reconstitution with water. It is sufficiently active and has an advantage over other available enzymes because it produces a negligible blank. In working with Sigma's firefly extract, the light–producing substances tend to precipitate and must be mixed prior to its use. Failure to do this, of course, will result in lack of reproducibility of the result. Therefore, it is recommended that the extract be thoroughly and gently mixed after reconstitution and centrifuged at 5,000 × g in cold (0 to 4°C). The clear supernatant, although not as active as the suspension, lends itself to accurate and reproducible work.

The presence of myokinase in the reaction mixture as well as in the firefly extract (De Leo and Giovannozzi-Sermanni, 1968) causes some recyclization of ADP to ATP which results in unexpectedly higher ATP values.

This complication may be avoided in one of three ways: (1) by removal of the interfering myokinase using the method of De Leo and Giovannozzi-Sermanni, which is a tedious procedure and results in loss of luciferase activity; (2) by inactivation of myokinase through addition of 20 to 40 μl of 6% H_2O_2 which is known to inhibit myokinase (Long, 1961) or (3) by making the measurement rapidly, since the response of the firefly enzyme to added ATP is essentially instantaneous and peaks in less than a second (Strehler, 1968), and the relatively slow myokinase does not substantially affect the accuracy of ATP determination if the measurements are made rapidly.

The presence of pyrophosphatase can convert adenyloxyluciferin to ATP and oxidized luciferin which results in higher values of ATP (see Breckenridge, 1965, and Strehler, 1968, for comprehensive reviews). Therefore, it is recommended that the purified phosphodiesterase be essentially free from pyrophosphatase (see part C, under Methodology).

Finally, it is important that the ATP concentration lies within the linear portion of the luciferase's saturation curve. This range depends on the volume of reaction mixture and includes concentration up to about 35 μg of ATP/cc solution (Strehler, 1968).

IV. DISCUSSION

A. Reproducibility, Recovery, and Specificity

When the amounts of cerebellar tissues were increased, the level of cyclic AMP increased linearly (Fig. 5). As a test of the reproducibility of the method, cerebral concentrations of cyclic AMP were determined in nine male rats. The mean ($\pm$ SEM) value was 48 $\pm$ 2 pmoles of protein, with a range in values of 38–60 pmoles/mg of protein. About 85% of the added cyclic AMP could be recovered from varying amounts of cerebellar tissue (Table 1). Firefly luciferase is specific for determining ATP. Luciferin, Mg^{++}, and ATP are the only known requirements. No known phosphorylated compound has been found to replace ATP; thus AMP, ADP, GDP, GTP, IDP, ITP, CDP, and CTP are inactive (Seliger and McElroy, 1965).

B. Modification of the Method

When determination of cyclic AMP in a whole organ such as the brain, liver, or kidney is desired, the entire tissue may be homogenized in a proportionally larger amount of $ZnSO_4$ (part F, Methodology), and the volume of $Ba(OH)_2$ adjusted accordingly. When dealing with larger amounts of tissue and consequently higher concentrations of cyclic AMP, the lyophilization step may be eliminated and the column eluant containing cyclic AMP divided into two equal parts representing the test and the blank. To eliminate this step, however, the concentration of "conversion" buffer must be increased

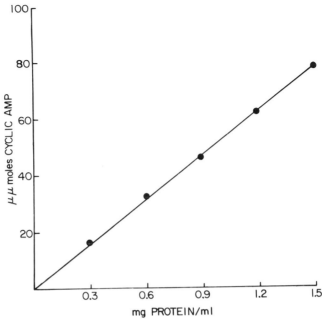

FIG. 5. The relationship between tissue levels of cyclic AMP and increasing amounts of rat cerebellar cortex. The reaction mixture contained cyclic AMP isolated from cerebellar homogenate (0.3–1.5 mg of protein), 500 μl of conversion buffer, and 5 μl of an enzyme mixture containing 2.5 μg of phosphodiesterase, 15 μg of myokinase, and 40 μg of pyruvate kinase. The reactions were carried out for 6 hr and were terminated by addition of 20 μl of 6% (v/v) H_2O_2. The samples were then frozen, thawed, and assayed for ATP using 500 μg of firefly luciferin-luciferase (part H, under Methodology). (Reprinted with permission from Ebadi et al., 1971a.)

three times (part E, Methodology). In this case, in determining cyclic AMP, 500 μl of the concentrated "conversion" buffer was added to 1.50 cc column eluant along with three times more of phosphodiesterase, myokinase, and pyruvate kinase than was recommended (part G, Methodology). The experiment was carried out as described and the resulting ATP was determined as described with the exception that the amount of luciferin–luciferase will be increased to 900 μg/ml.

On the other hand, when the lack of sufficient tissue (e.g., rat pineal gland) required a microassay, the following modification was instituted. The tissues were homogenized in 300 μl of 0.25 M $ZnSO_4$ without addition of labeled cyclic AMP. The isolation, purification, and conversion of cyclic AMP to ATP were carried out as usual. However, the concentration of H_2O_2 was reduced to 5 μl of the 6% (v/v) solution, the reaction mixture was thoroughly mixed, transferred to a culture tube (0.5 × 5 cm, A.H. Thomas Co., Philadelphia, Pa.), and freeze-dried. The blanks lacking cyclic AMP but containing the buffer, enzymes, and H_2O_2 were similarly treated. The ATP was then dissolved in 90 μl of 50 mM Tris-HCl (pH 7.5), containing 3 mM $MgSO_4$, and the culture tube was placed inside a scintillation counting vial. The reaction was started by mixing 100 μg of firefly extract previously placed against the wall of the tube a few mm above the reaction mixture. Light emission was determined as previously described. With this technique, concentrations of cyclic AMP standards as low as 5×10^{-14} M carried through the entire procedure were easily detected, with the test values reading approximately 10 times higher than the blank values (1,200 cpm vs 125 cpm).

C. Subcellular and Tissue Levels of Cyclic AMP

The regional and subcellular distribution of cyclic AMP in the brain is shown in Tables 2 and 3. Virtually all of the cyclic AMP is associated with particulate material in the nuclear and mitochondrial fractions. Practically no cyclic AMP was present in the soluble, high-speed supernatant (Ebadi et al., 1971b). In the brain, the highest concentration of cyclic AMP was found in the pineal gland, cerebellum, pituitary, and spinal cord (Ebadi et al., 1971b). The concentration of cyclic AMP depends and reflects upon the relative activity of adenyl cyclase and phosphodiesterase. The levels of cyclic AMP in various rat tissues have been listed in Table 4 in descending order of concentrations. In general, the active organs such as the liver, heart, and kidney, and glandular tissues such as thyroid and adrenal glands are richly endowed with cyclic AMP. These levels of cyclic AMP are somewhat greater than those

TABLE 2. *Subcellular distribution of cyclic AMP in rat brain[a]*

Fraction	Cyclic AMP		Whole homogenate (%)
	Protein (pmoles/mg)	pmoles/ fraction	
whole homogenate	38 ± 5	5,880 ± 710	100
nuclear	44 ± 7	1,830 ± 220	31
mitochondrial	59 ± 6	2,030 ± 130	35
microsomal	17 ± 9	55 ± 26	1
soluble	16 ± 5	160 ± 50	3

[a]Each figure represents the mean value of six experiments ± SE.

Male Sprague-Dawley rats were decapitated and the brains were immediately homogenized in 0.32 M sucrose containing 0.28 M $ZnSO_4$. The subcellular fractions were obtained by differential centrifugation according to the method of De Robertis et al. (1967). Cyclic AMP was isolated, converted to ATP, and determined with the luciferin-luciferase system (Methodology). (Reprinted with permission from Ebadi et al., 1971b.)

TABLE 3. *Regional distribution of cyclic AMP in rat brain*

Brain area	Cyclic AMP ($\mu\mu$moles/mg protein)
pineal gland	58 ± 7 (8)
cerebellum	48 ± 2 (8)
pituitary gland	35 ± 3 (8)
spinal cord	25 ± 2 (8)
thalamus	22 ± 2 (8)
telencephalon	20 ± 2 (6)
hypothalamus	19 ± 1 (8)
olfactory bulb	19 ± 2 (6)
inferior colliculus	15 ± 2 (7)
superior colliculus	14 ± 2 (7)
olfactory tubercle	14 ± 2 (7)
caudate nucleus	14 ± 2 (7)
hippocampus	13 ± 1 (7)
pons	11 ± 2 (6)
medulla	11 ± 2 (6)

Male Sprague-Dawley rats were decapitated and the brain areas were dissected according to an atlas by König and Klippel (1963). The level of cyclic AMP changes with time after decapitation and this alteration proceeds at different rates in different structures. Therefore, various brain areas were isolated and kept on aluminum foil and were frozen on dry ice in exactly one minute. Cyclic AMP was isolated, converted to ATP, and determined by luciferin-luciferase (Methodology). (Reprinted with permission from Ebadi et al., 1971*b*.)

previously reported (Posner, Hammermeister, Bratvold, and Krebs, 1964; Breckenridge, 1965; Butcher, Ho, Meng, and Sutherland, 1965; Pauk and Reddy, 1967; Turtle and Kipnis, 1967; Brooker, Thomas, and Appelman, 1968; Goldberg et al., 1969; Steiner, Kipnis, Utiger, and Parker, 1969). Variations in the tissue levels of cyclic AMP reported by different investigators may be explained by the rapid postmortem changes in the level of cyclic AMP (Ebadi et al., 1971*a*). In this case, all tissues were excised and frozen in exactly 1 min.

D. Estimation of Cyclic AMP in the Urine

Randomly voided and 24-hr collected urine samples were kept frozen until assayed. The insoluble substances were removed by either centrifugation

TABLE 4. *Contents of cyclic AMP in various organs of the rat*

Tissue	Cyclic AMP (nmole/g wet wt. of tissue)
stomach (pyloric region)	4.2 ± 0.4 (3)
liver	3.3 ± 0.2 (6)
adrenal (whole)	3.2 ± 0.6 (3)
kidney	3.0 ± 0.2 (10)
thyroid (whole)	2.9 (2)
heart (left ventricle)	2.7 ± 0.4 (7)
pancreas	2.3 ± 0.4 (3)
muscle (gastroenemius)	1.7 ± 0.3 (5)
spleen	1.6 ± 0.2 (6)
submaxillary gland	1.6 ± 0.2 (4)
skin (abdominal)	1.5 ± 0.3 (4)
small intestine (duodenum)	1.5 ± 0.2 (5)
aorta (ascending)	1.5 (2)
trachea (bifurcation)	1.5 ± 0.3 (3)
testes (whole)	1.5 ± 0.2 (6)
aorta (descending)	1.3 ± 0.5 (3)
large intestine (descending colon)	1.3 ± 0.1 (6)
thymus (adult)	1.2 ± 0.3 (3)
lung (right superior lobe)	1.0 ± 0.2 (5)
stomach (greater curvature)	1.0 ± 0.2 (4)
adipose tissue	0.35 ± 0.10 (4)

The level of cyclic AMP in an isolated organ changes with time and this alteration proceeds at different rates in different structures. Therefore, after decapitation of the rats, different organs were removed and frozen in exactly one minute. Cyclic AMP was isolated, converted to ATP, and determined by luciferin-luciferase (Methodology). (Reprinted with permission from Ebadi et al., 1971a.)

or filtration by ordinary filter paper. Although no 5'-nucleotides occur in the urine (Ashman, Lipton, Melicow, and Price, 1963; Price, Ashman, and Melicow, 1967), the $ZnSO_4$–$Ba(OH)_2$ treatment was carried out as usual by using equal volumes of urine, $ZnSO_4$, and $Ba(OH)_2$. After precipitation, the supernatant was diluted 1:100 with distilled water. A 0.5 cc portion of the diluted sample was chromatographed on Dowex-50, and the concentration of eluted cyclic AMP was assayed as described (parts F–H, under Methodology) and its concentration reported as μmoles/24-hr excretion. The 24-hr urinary excretion in normal humans without any dietary restriction was shown to be 7.5 ± 0.60 μmoles.

V. SUMMARY

A simple, sensitive, and specific method for assaying cyclic AMP in various tissues is reported. Cyclic AMP was isolated from contaminating nucleotides and was converted to ATP with a phosphodiesterase–myokinase–pyruvate kinase system. The ATP was determined enzymically in a liquid scintillation counter by the firefly luciferin–luciferase technique. This procedure was capable of detecting as little as 5×10^{-14} moles of cyclic AMP and could therefore be used for analyses on less than 1 mg of brain. The assay was reproducible and linear over a wide range of tissue concentrations.

In the rat, the highest levels of cyclic AMP (2.7 to 4.2 pmoles/mg wet wt. of tissue) were present in the heart, thyroid, kidney, adrenal, liver, and pyloric region of the stomach. Intermediate levels (1.5 to 2.7 pmoles/mg wet wt. of tissue) were found in testis, skin, aorta, intestine, submaxillary gland, spleen, muscle, and cerebral cortex. Moderately low levels (1.0 to 1.5 pmoles/mg wet wt. of tissue) were found in lung, trachea, and greater curvature of the stomach, whereas low levels (0.15 to 0.60 pmoles/mg wet wt. of tissue) were found in adipose tissue. The 24-hr urinary excretion in normal humans without dietary restriction was shown to be $7.5 \pm .60$ μmoles.

Studies of the subcellular distribution of cyclic AMP showed that virtually all the cyclic AMP was bound to particulate material associated with the nuclear and mitochondrial fractions of brain homogenates. In studying the regional distribution of cyclic AMP in the brain, the highest concentrations (35 to 58 $\mu\mu$moles/mg protein) were found in the pituitary and pineal glands and in the cerebellum; intermediate levels (19 to 25 $\mu\mu$moles/mg protein) were found in spinal cord, thalamus, hypothalamus, telencephalon, and olfactory bulb; the lowest concentrations (11 to 15 $\mu\mu$moles/mg protein) were found in corpora quadrigemina, olfactory tubercle, caudate, hippocampus, pons, and medulla.

Acknowledgment. The authors gratefully acknowledge the editorial and secretarial assistance of Mrs. Sue Eulberg and Mrs. Bev Grenier. This study was supported in part by U.S. Public Health Service grants HD 00370 from the National Institute of Child Health and Human Development and NS 08932 from the National Institute of Neurological Disease and Stroke.

REFERENCES

Addanki, G., Sotos, J. F., and Rearick, D. (1966): Rapid determination of picomole quantities of ATP with a liquid scintillation counter. *Analytical Biochemistry*, 14:261–264.

Ashman, D. F., Lipton, R., Melicow, M. M., and Price, T. D. (1963): Isolation of adenosine 3′,5′-monophosphate and guanosine 3′,5′-monophosphate from rat urine. *Biochemical Biophysical Research Communications*, 11:330–334.

Balharry, G. J. E., and Nicholas, D. J. D. (1971): New assay for ATP-sulphurylase using the luciferin-luciferase method. *Analytical Biochemistry*, 40:1–17.

Bray, G. A. (1960): A simple efficient liquid scintillator for counting aqueous solutions in a liquid scintillation counter. *Analytical Biochemistry*, 1:279–285.

Breckenridge, B. M. (1965): The measurement of cyclic adenylate in tissues. *Proceedings of the National Academy of Sciences*, 52:1580–1586.

Brooker, G., Thomas, L. J., and Appelman, M. M. (1968): The assay of adenosine 3',5'-cyclic monophosphate and guanosine 3',5'-cyclic monophosphate in biological materials by enzymatic radioisotopic displacement. *Biochemistry*, 7:4177–4181.

Butcher, R. W., and Sutherland, E. W. (1962): Adenosine 3',5'-phosphate in biological materials. I. Purification and properties of cyclic 3',5'-nucleotide phosphodiesterase and use of this enzyme to characterize adenosine 3',5'-phosphate in human urine. *Journal of Biological Chemistry*, 237:1244–1250.

Butcher, R. W., Ho, R. J., Meng, H. C., and Sutherland, E. W. (1965): Adenosine 3',5'-monophosphate in biological materials. II. The measurement of adenosine 3',5'-monophosphate in tissues and the role of the cyclic nucleotide in the lipolytic response to fat to epinephrine. *Journal of Biological Chemistry*, 240:4515–4523.

Chan, P. S., Black, C. T. and Williams, B. J. (1970): Separation of cyclic 3',5'-AMP (CAMP) from ATP, ADP, and 5'-AMP by precipitation in adenyl cyclase assay. *Federation Proceedings*, 29:2080.

Chase, A. M. (1960): The measurement of luciferin and luciferase. In: *Methods of Biochemical Analysis*, Vol. 8, edited by D. Glick. Interscience Publishers, New York, 61–117.

Cook, W. H., Lipkin, D., and Markham, R. (1957): The formation of a cyclic dianhydrodiadenylic acid (I) by the alkaline degradation of adenosine-5'-triphosphoric acid (II). *Journal of American Chemical Society*, 79:3607–3608.

De Leo, P., and Giovannozzi-Sermanni (1968): Interference of myokinase in the determination of ATP by luciferase assay. *Biochimica et Biophysica Acta*, 170:208–210.

De Robertis, E., Rodriguez De Lores, Arn., Alberici, M., Butcher, R. W., and Sutherland, E. W. (1967): Subcellular distribution of adenyl cyclase and cyclic phosphodiesterase in rat brain cortex. *Journal of Biological Chemistry*, 242:3487–3493.

Ebadi, M. S.: The measurement of cyclic nucleotides and the related enzymes by luciferin-luciferase system. *Methods in Neurochemistry*, 3 (*in press*).

Ebadi, M. S., Weiss, B., and Costa, E. (1971a): Microassay of adenosine-3',5'-monophosphate (cyclic AMP) in brain and other tissues by the luciferin-luciferase system. *Journal of Neurochemistry*, 18:183–192.

Ebadi, M. S., Weiss, B., and Costa, E. (1971b): Distribution of cyclic adenosine monophosphate in rat brain. *Archives of Neurology*, 24:353–357.

Goldberg, N. D., Larner, J., Sasko, H., and O'Toole, A. G. (1969): Enzymatic analysis of cyclic 3',5'-AMP in mammalian tissues and urine. *Analytical Biochemistry*, 28:523–544.

Holmsen, H., Holmsen, I., and Bernhardsen, A. (1966): Microdetermination of adenosine diphosphate and adenosine triphosphate in plasma with the firefly luciferase system. *Analytical Biochemistry*, 17:456–473.

Johnson, R. A., Hardman, J. G., Broadus, A. E., and Sutherland, E. W. (1970): Analysis of adenosine 3',5'-monophosphate with luciferase luminescence. *Analytical Biochemistry*, 35:91–97.

König, J. F. R., and Klippel, R. A. (1963): *The Rat Brain*. William and Wilkins Company, Baltimore.

Krishna, G., Weiss, B., and Brodie, B. B. (1968): A simple, sensitive method for the assay of adenyl cyclase. *Journal of Pharmacology and Experimental Therapeutics*, 163:379–385.

Krishna, G., and Birnbaumer, L. (1970): On the assay of adenyl cyclase. *Analytical Biochemistry*, 35:393–397.

Long, C. (1961): Adenylate kinase (myokinase). In: *Biochemists' Handbook*, edited by C. Long. E. and F. N. Spon Ltd., London, p. 410.

Lowry, O. H., Rosebrough, N. J., Farr, A. L., and Randall, R. L. (1951): Protein measurement with the folin phenol reagent. *Journal of Biological Chemistry*, 193:265–275.

Lyman, G. E., and DeVincenzo, J. P. (1967): Determination of picogram amounts of ATP using the luciferin-luciferase enzyme system. *Analytical Biochemistry*, 21:435–443.

McElroy, W. D. (1947): The energy source for bioluminescence in an isolated system. *Proceedings of the National Academy of Sciences*, 33:342–345.

McElroy, W. D., and Strehler, B. L. (1949): Factors influencing the response of the bioluminescent reaction to adenosine triphosphate. *Archives of Biochemistry*, 21–22:420–433.

McElroy, W. D., and Seliger, H. H. (1961): Mechanism of bioluminescent reactions. In: *A Symposium on Light and Life*, edited by W. D. McElroy and B. Glass. The Johns Hopkins Press, Baltimore, pp. 219–257.

Nair, K. G. (1966): Purification and properties of 3',5'-cyclic nucleotide phosphodiesterase from dog heart. *Biochemistry*, 5:150–157.

Pauk, G. L., and Reddy, W. J. (1967): Measurement of adenosine 3',5'-monophosphate. *Analytical Biochemistry*, 21:298–307.

Posner, J. B., Hammermeister, K. E., Bratvold, G. E., and Krebs, E. G. (1964): The assay of adenosine 3',5'-phosphate in skeletal muscle. *Biochemistry*, 3:1040–1044.

Price, T. D., Ashman, D. F., and Melicow, M. M. (1967): Organophosphates of urine, including adenosine 3',5'-monophosphate and guanosine 3',5'-monophosphate. *Biochimica et Biophysica Acta*, 138:452–465.

St. John, J. B. (1970): Determination of ATP in chlorella with the luciferin-luciferase enzyme system. *Analytical Biochemistry*, 37:409–416.

Seliger, H. H., and McElroy, W. D. (1965): Bioluminescence enzyme catalyzed chemiluminescence. In: *Light: Physical and Biological Action*, edited by H. H. Seliger and W. D. McElroy. Academic Press, New York, pp. 168–205.

Stanley, P. E., and Williams, S. G. (1969): Use of liquid scintillation spectrometer for determining adenosine triphosphate by the luciferase enzyme. *Analytical Biochemistry*, 29:381–392.

Stanley, P. E. (1971): Determination of subpicomole levels of NADH and FMN using bacterial luciferase and the liquid scintillation spectrometer. *Analytical Biochemistry*, 39:441–453.

Steiner, A. L., Kipnis, D. M., Utiger, R., and Parker, C. (1969): Radioimmunoassay for the measurement of adenosine 3',5'-cyclic phosphate. *Proceedings of the National Academy of Sciences*, 64:367–373.

Strehler, B. L., and Totter, J. R. (1952): Firefly luminescence in the study of energy transfer mechanisms. I. Substrate and enzyme determination. *Archives of Biochemistry and Biophysics*, 37–40:28–41.

Strehler, B. L. (1968): Bioluminescence assay: principles and practice. In: *Methods of Biochemical Analysis*, Vol. 16, edited by D. Glick. Interscience Publishers, New York, pp. 99–181.

Sutherland, E. W., Rall, T. W., and Menon, T. (1962): Adenyl cyclase. I. Distribution, preparation and properties. *Journal of Biological Chemistry*, 237:1220–1227.

Tal, E., Dikstein, S., and Sulman, F. G. (1964): ATP determination with the tricarb scintillation counter. *Experientia*, 20:652.

Turtle, J. R., and Kipnis, D. M. (1967): A new assay for adenosine 3',5'-cyclic monophosphate in tissue. *Biochemistry*, 6:3970–3976.

Weiss, B. (1971): A rapid microassay of phosphodiesterase activity. *Transactions of the American Society for Neurochemistry*, 2:117.

Yamamoto, M., and Massey, K. L. (1969): Cyclic 3',5'-nucleotide phosphodiesterase of fish (Salmo gairdnerii) brain. *Comparative Biochemical Physiology*, 30:941–954.

EDITORIAL NOTE ADDED IN PROOF

Dr. E. Costa has advised us that he and his colleagues have modified the method described here by Dr. Ebadi, thus eliminating interference by an unknown contaminant which appears to have been responsible for the unusually high levels of cyclic AMP previously reported.

In brief summary, frozen tissues (25 to 100 mg) were homogenized in 500 μl of cold 0.4 N perchloric acid containing ^{3}H-cyclic AMP (2×10^{-14} moles, specific activity 3 C/mmole) to serve as an internal standard and correct for the losses during homogenization and purification of the tissue extract. Solutions of authentic cyclic AMP (2.5–1,000 pmoles) were carried through the method in parallel with tissue samples. Potassium bicarbonate (0.4 M) was added, and the homogenate was adjusted to pH 7.5 with 0.6 M Tris buffer. After centrifugation, 1 ml of the supernatant fluid was placed on a neutral alumina column (4.5×0.4 cm, equilibrated with 0.06 M Tris buffer, pH 7.5). The effluent was discarded, and the cyclic AMP was eluted with 2 ml of 0.06 M Tris buffer (pH 7.5). This eluate was drained directly onto a cation-exchange column (AG 50WX8, 200–400 mesh, H$^+$, 3.5×0.4 cm). The effluent of the cation-exchange column was discarded, and the column was then eluted with water. The first 3 ml was discarded, and the following 2 ml was collected: this portion contained 80 to 90% of the ^{3}H-cyclic AMP and was free of other known nucleosides and nucleotides.

Dr. Costa has further advised us that this revised method has been successfully applied to the measurement of cyclic AMP in the mouse parotid gland, as described in a forthcoming paper (A. Guidotti, B. Weiss, and E. Costa, *Molec. Pharmacol.*, in press). The data incidentally support the hypothesis that isoproterenol-induced DNA synthesis in the parotid gland is mediated by cyclic AMP.

—P.G., R.P., and G.A.R.

Advances in Cyclic Nucleotide Research, Vol. 2
Raven Press, New York © 1972

High-Pressure Anion Exchange Chromatography and Enzymatic Isotope Displacement Assays for Cyclic AMP and Cyclic GMP

Gary Brooker*

Departments of Medicine and Biochemistry, University of Southern California, School of Medicine, Los Angeles, California 90033

I. INTRODUCTION

A monumental amount of information about cyclic AMP has now been obtained, mainly by Sutherland and his collaborators. There are several excellent review articles on this subject (Sutherland and Rall, 1960; Robison, Butcher, and Sutherland, 1968). The original assay system for cyclic AMP stemmed from Rall and Sutherland's discovery that cyclic AMP was the heat-stable factor, formed in response to epinephrine, which activated liver glycogen phosphorylase (Rall and Sutherland, 1958).

A large number of alternative assay methods for cyclic AMP have now been developed. A recent review of assay methods has been made (Breckenridge, 1971). I would like first to review our enzymatic isotope displacement assay (Brooker, Thomas, and Appleman, 1968) and some modifications which have improved the speed, sensitivity, and reliability of the assay. Secondly, I would like to describe a new physical method for cyclic AMP analysis based upon high-pressure anion exchange chromatography (Brooker, 1971a).

---

* Present address: Department of Pharmacology, University of Virginia School of Medicine, Charlottesville, Virginia 22901.

II. ENZYMATIC RADIOISOTOPE DISPLACEMENT ASSAY OF CYCLIC AMP

A. Theory

The principle of enzymatic isotope displacement analysis developed by Brooker et al. (1968), Brooker and Appleman (1968), and Newsholme and Taylor (1968) has been applied to the assay of cyclic nucleotides. If a radioactive substrate S* of high specific activity is allowed to partially react in an enzymatic reaction, addition of nonradioactive substrate S will reduce the formation of radioactive product (P*). The ratio of this reduction, P^*/P_0^* (radioactive product in the presence of radioactive S* and nonradioactive substrate S/radioactive product in the presence of only radioactive substrate S*), or dilution in radioactivity by a certain concentration of nonradioactive substrate S, is a function of the concentration of radioactive substrate, and the Michaelis

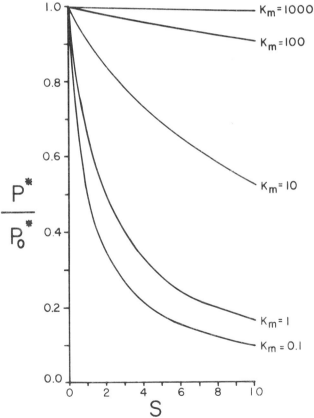

FIG. 1. Plot of P*, P_0^* vs S, values calculated from $P^*/P_0^* = (K_m + S^*)/(K_m + S + S^*)$. The value S* is held constant at 1 while K_m varies as indicated. (From Brooker and Appleman, 1968)

constant (K_m) of the enzyme used. The following equation (1) from Brooker et al. (1968) defines this relationship.

$$\frac{P^*}{P_0^*} = \frac{K_m + S^*}{K_m + S + S^*} \tag{1}$$

Figures 1 and 2 graphically depict this relationship. It can be seen that the greatest sensitivity is obtained when the K_m and radioactive substrate concentration S^* are less than the concentration of nonradioactive substrate S being measured.

This assay principle was used to measure cyclic AMP using a phosphodiesterase preparation from rat brain. We found that a low K_m form (1×10^{-6} M) of the enzyme existed (Brooker et al., 1968) along with the high K_m form (1.3×10^{-4} M) as originally described by Butcher and Sutherland (1962). Kinetic analysis showed two enzyme activities existed in our rat brain phospho-

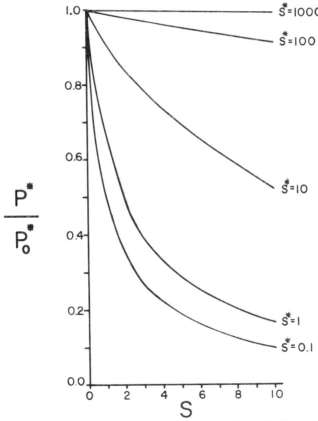

FIG. 2. Plot of P^*/P_0^* vs S, values calculated from $P^*/P_0^* = (K_m + S^*)/(K_m + S + S^*)$. The value of K_m is held constant at 1 while S^* varies as indicated. (From Brooker and Appleman, 1968)

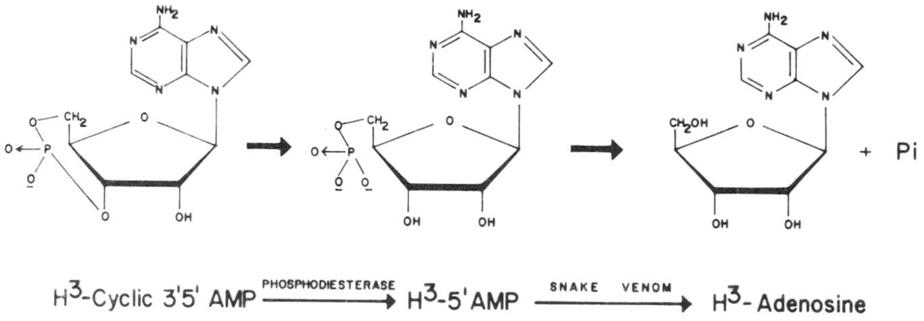

H^3-Cyclic 3'5' AMP $\xrightarrow{\text{PHOSPHODIESTERASE}}$ H^3-5'AMP $\xrightarrow{\text{SNAKE VENOM}}$ H^3-Adenosine

FIG. 3. Scheme for enzymatic isotope displacement assay of cyclic AMP. (From Brooker et al., 1968)

diesterase preparation. One appeared specific for cyclic GMP, and the other specific for cyclic AMP. Thompson and Appleman (1971) have now separated these enzymes and have suggested that the high k_m cyclic AMP phosphodiesterase is actually the cyclic GMP specific phosphodiesterase. The use of the simple rat brain phosphodiesterase preparation with apparent Michaelis constants of 1×10^{-6} M for cyclic AMP and 5×10^{-6} M for cyclic GMP has yielded simple assays for cyclic AMP and cyclic GMP.

As shown in Fig. 3 (Scheme I from Brooker et al., 1968), introduction of nonradioactive cyclic AMP will reduce the amount of ^{3}H 5'-AMP formed. The complete reaction is carried out in a liquid scintillation vial with an excess of 5' nucleotidase activity present to convert all 5' AMP formed to adenosine. The amount of phosphodiesterase is chosen to convert only about 40% of ^{3}H-cyclic AMP to ^{3}H-5'-AMP. The reaction is terminated by addition of anion exchange resin, which binds the unreacted ^{3}H-cyclic AMP substrate, while the product of the reaction, ^{3}H-adenosine, is not bound. Scintillation fluid (P-dioxane) is added to the vial and the sample counted in a liquid scintillation counter. The unreacted substrate which is bound to the resin is quenched, while the product, ^{3}H-adenosine, is neither bound nor quenched and is detected in the liquid scintillation process.

B. Reagents

Tritiated cyclic AMP (16.3 C/mMole), cyclic AMP, and 8-^{14}C-adenosine (58.3 mC/mMole) were obtained from Schwarz BioResearch (Orangeburg, N.J.). Nucleotides and nucleosides were purchased from Sigma Chemical Co. (St. Louis, Mo.). Bovine serum albumin and king cobra (*Ophiophagus hanna*) snake venom were obtained from Sigma. AG-1-X2 50-100 and -400 mesh anion exchange resin was purchased from Bio Rad Laboratories (Richmond, Calif.). It was processed as previously described (Brooker et al., 1968). Precoated cellulose thin layer chromatography plates were Baker Flex No. 4468, obtained from the J. T. Baker Chemical Co. (Phillipsburg, N.J.). Scintillation solvent consisted of 22.7 g 2,5-diphenyloxazole and 125 g naphthalene per liter of P-dioxane (Eastman No. 2144). Polyethylene scintillation vials were used with 10 ml of scintillant. All other reagents were ACS reagent quality or better. All water used in these experiments was either distilled deionized or 18 megohm/cm deionized water. Disposable polypropylene (17 mm × 100 mm) test tubes, obtained from Falcon Plastics (Los Angeles, Calif.), were used for the homogenization and processing of tissue and urine samples prior to chromatography.

C. Preparation of Cyclic AMP Phosphodiesterase

A rat brain, about 1.7 g, was obtained by decapitation of a male 150 g Sprague-Dawley rat. The brain was homogenized with a Teflon and glass homogenizer (A. H. Thomas No. C22225) at 0 to 1°C for 1 min in 10 volumes of water per gram brain. The supernatant solution after centrifugation for 10 min at 30,000 × g (tip) was mixed with 200 mg/g original weight of activated charcoal (Matheson CX 648) and recentrifuged for 10 min at the same speed. The supernatant solution was withdrawn and stored at 4°C.

D. Enzyme Displacement Assay Procedure

To achieve greater sensitivity, the reaction volume has been reduced to 35 μl and higher specific activity ^{3}H-cyclic AMP used. In addition, an improved method for purification of cyclic AMP from tissue has been developed. The tissue prepurification method for the enzymatic isotope displacement assay is essentially the same as that used for the high-pressure anion exchange chromatography method. The unknown sample can be assayed with or without the thin-layer chromatography step. The evaporated sample is usually dissolved in 50 μl of 100 mM Tris-HCl, pH 8.0, and two 20 μl fractions assayed. The remaining 10 μl is used to determine purification recovery.

The cyclic AMP standards or unknown samples in 20 μl are added to the liquid scintillation vials. Five microliters of a 100 mM Tris-HCl, pH 8.0, solution containing 0.1 μC ^{3}H cyclic AMP, 1.2 μmoles 5′-AMP, 24 μmoles EGTA [ethylenebis (oxyethylenenitrilo) tetraacetic acid], and 1.2 μmoles MgCl$_2$ is placed adjacent to the standards or unknown samples. The reaction is then started by placing 10 μl of an enzyme mixture in 100 mM Tris-HCl, pH 8.0, containing 0.1 to 0.3 μl brain phosphodiesterase, and 10 μg each of bovine serum albumin and king cobra venom nucleotidase. It is important to premix the phosphodiesterase snake venom solution just before commencing the incubation. The vials are each incubated for 10 min and the reaction stopped by addition of 0.8 ml of a 50% slurry of AG-1-X2-400 mesh anion exchange resin. After the mixture has equilibrated for 10 min, 10 ml of scintillation fluid is added and the samples counted. A typical standard curve is shown in Fig. 4. The values for unknown samples reacted in the same way are obtained by reference to the standard curve.

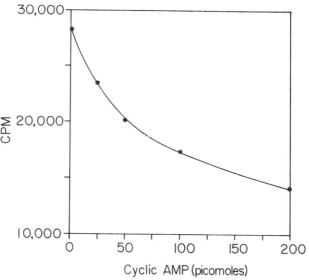

FIG. 4. Standard curve for measurement of cyclic AMP by improved enzymatic isotope displacement assay.

III. HIGH-PRESSURE ANION EXCHANGE CHROMATOGRAPHIC MEASUREMENT OF CYCLIC AMP

A. Introduction

High-pressure anion exchange chromatography offers several advantages over conventional chromatographic methods. The most important is the ability to elute the sample in a small volume for detection by an ultrasensitive 254 nm absorption flow cell (Brooker, 1971*b*). The presence of 1 pmole of cyclic AMP within the 8 μl flow cell causes a full scale deflection on the strip chart recorder. Another advantage of this system is that high-speed chromatographic separations and measurements can be made. Typically, a single determination takes less than 8 min. This high-pressure chromatographic system was first shown by Brooker (1970) to be applicable to measure cyclic AMP produced from rat brain adenylate cyclase. More recently, methods have been developed which permit measurement of cyclic AMP in many types of biological materials (Brooker, 1971*a,c*). In addition, ^{14}C-cyclic AMP specific activity can be determined in tissues prelabeled with ^{14}C-adenine or -adenosine.

B. Apparatus

High-pressure anion exchange chromatography was performed using a 3-m capillary "pellicular" anion exchange column as originally described by Horvath, Preiss, and Lipsky (1967). A Varian-Aerograph LCS-1000 liquid chromatograph with an 8 μl, 1 cm pathlength, 254 mμ ultraviolet flow cell was used. The detector was modified as previously described (Brooker, 1971b), and the resin in the instrument's precolumn was removed to allow injection of up to 30 μl samples. A Varian-Aerograph 1 mV full-scale strip recorder was used. Full-scale deflection of the recorder equalled 0.008, 0.004, or 0.002 absorbancy units as indicated in the text. The resin in the 3-m column was converted to the chloride form by washing the column with 0.1 N HCl, and then to neutrality with water. Cyclic AMP was extracted from tissues by homogenization with a polytron Model PT10 purchased from Brinkman Instruments (Westbury, N.Y.). Standards of cyclic AMP were made in a Beckman DB Spectrophotometer using extinction coefficients given by Smith, Drummond, and Khorana (1961). Liquid scintillation counting was performed on a three channel Beckman LS250 liquid scintillation counter with AQC. Counting efficiency for tritium was 38% and 60% for ^{14}C. The ^{14}C channel was adjusted so that no tritium counts occurred within it. Carbon-^{14} efficiency in the tritium channel was 6%.

C. High-Pressure Anion Exchange Chromatography and Quantitation of Cyclic AMP

The chromatograph was operated in a nongradient mode with HCl, pH 2.20 as the eluting buffer. This allowed the continuous injection of one

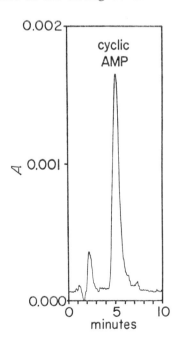

FIG. 5. High-pressure anion exchange chromatographic measurement of cyclic AMP. Cyclic AMP peak is 43 pmoles. (From Brooker, 1971b)

sample after another without the necessity of column regeneration. Injection of 43 pmoles of cyclic AMP into the chromatograph gives nearly a full scale peak on the strip chart recorder, as shown in Fig. 5. Under these conditions, the instrument response is linear from 0 to 100 pmoles of cyclic AMP simply by measuring the peak height of the chromatogram (Brooker, 1970), as shown in Fig. 6. For tissue analysis, full-scale deflection equalled 0.002 absorbancy units and flow was 24 ml/hr. The column temperature was 80°C. Inlet pressure was 1,050 to 1,200 psi at 24 ml/hr. Samples were injected in volumes of 25 μl or less, the system closed, and the flow started. A time-actuated automatic fraction collector was used to collect the cyclic AMP peak in a liquid scintillation vial. Measurement of the peak height of the unknown and comparison to the peak height of the standard was used to determine the amount of nonradioactive cyclic AMP in the sample. Measurement of tritium radioactivity in the cyclic AMP peak was used to determine the percentage of the original sample that was measured. Division of the nonradioactive cyclic AMP by the fractional isotopic recovery equalled the amount of cyclic AMP present in the original tissue extract or urine specimen. ^{14}C cyclic AMP specific activity was also equal to the quotient of the ^{14}C radioactivity in the total cyclic AMP peak (corrected for recovery) divided by the amount of cyclic AMP in the original sample (corrected for recovery).

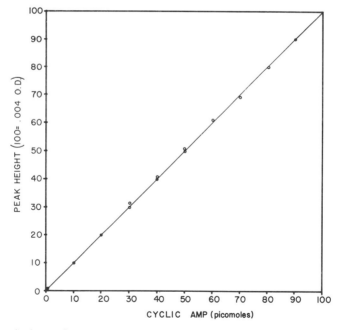

FIG. 6. Standard curve for measurement of cyclic AMP by the high-pressure chromatographic system. (From Brooker, 1970)

D. Prepurification of Cyclic AMP from Tissue

Ten to 400 mg of frozen tissue obtained by freezing between clamps precooled in liquid nitrogen was placed into a 17 × 100 mm polypropylene test tube precooled in liquid nitrogen. Ten microliters of ^{3}H-cyclic AMP tracer (0.1 to 1 pmoles) was added, and the tube was placed under the polytron homogenizer, 3 ml of 5% trichloroacetic acid added, and the sample quickly homogenized at full speed for 10 sec. The homogenate was centrifuged in a bench-top centrifuge and the supernatant was decanted to another identical tube. The trichloroacetic acid was extracted with three 8-ml extractions of water-saturated ethyl ether. After each addition of ether, the tube was mixed on a Variwhirl vortex mixer (Van Waters & Rogers No. 5810-006) for 10 sec. After the ether had separated, it was removed with a Teflon-tipped suction aspirator. Two-hundred microliters of 0.5 M Tris-HCl, pH 8.0, was added and the sample applied to an anion exchange column filled with AG-1-X2 50–100 mesh chloride form resin prepared in a disposable Pasteur pipet (0.8 × 5 cm). A small polypropylene funnel (A. H. Thomas Co. No. 5587G) was attached to the top of the column which acted as a solution reservoir above the column. The solution was allowed to pass through the column and 5 ml of water applied. After the water, 10 ml of HCl, pH 1.3, was applied and collected to elute cyclic AMP from the column and separate it from tissue extract ATP which remained on the column. It was necessary to eliminate ATP from the sample before treatment with Zn-Ba, to prevent the formation of nonenzymatically produced cyclic AMP from ATP (Brooker, 1971c). To the cyclic AMP fraction, 100 μl of 2 M Tris was added, followed by 2.5 ml of 5% $ZnSO_4$ and 2.5 ml of 0.3 N $Ba(OH)_2$. The precipitate was mixed and then centrifuged in a bench-top centrifuge. Ten milliliters of water was applied to the column to ready it for reapplication of the $ZnSO_4$–$Ba(OH)_2$ supernatant. A new column was used when tissue samples had been prelabeled with ^{14}C-adenosine to determine ^{14}C-cyclic AMP specific activity. If the samples were to be assayed by the enzyme displacement method, a sheet of Whatman No. 42 filter paper was placed on the solution reservoir funnel to prevent any colloidal Zn-Ba from contaminating the Dowex resin. Once the Zn-Ba supernatant had been applied to the column, the filter paper was removed. After the zinc-barium supernatant was applied to the column, it was followed by 30 ml of HCl, pH 2.7, followed by collection of 25 ml of HCl, pH 2.1, into a 40 ml conical centrifuge tube (Pyrex No. 8124). The flow rate for these columns was about 3 ml per min. The cyclic AMP fractions were evaporated to dryness under vacuum at 55°C on a Buchler Evapomixer. The samples were taken up and spotted on cellulose thin-layer plates to eliminate interfering substances which continuously bleed from the AG-1-X2 resin. Figure 7A shows the high pressure chromatogram from a reagent blank processed without thin–layer chromatography or the final ether extraction. Figure 7B shows the chromatograph from a complete reagent blank which includes

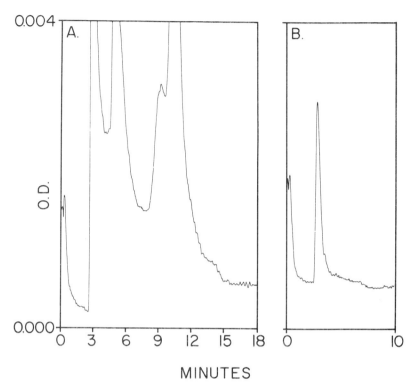

FIG. 7. A. High-pressure anion exchange chromatograph of reagent blank without thin-layer chromatography and ether extraction. B. Complete reagent blank with thin layer chromatography and ether extraction.

the thin-layer and ether extraction steps. It can be seen that thin-layer chromatography and ether extraction effectively eliminates the interfering substances.

Reference spots of cyclic AMP (1 μmole) were spotted at each edge of the thin-layer plates. Usually, 8 to 12 samples were spotted on a 20 × 20 cm plate. The plates were developed for a distance of 10 to 15 cm with n-butanol:acetic acid:water, 2:1:1. The reference spots of cyclic AMP, at each edge, were visualized with a short wavelength UV light, the area of each unknown sample cut from the plate, and the cellulose scraped into a polypropylene test tube. Two milliliters of water was added, the sample vigorously mixed for 15 sec, and the cellulose centrifuged. The supernatant was transferred to a 40-ml conical centrifuge tube and extracted three times with 8 ml of ether. The sample was evaporated to dryness on the Evapomixer and the sample taken up with 30 μl of HCl, pH 2.20, and 20 μl injected into the high-pressure chromatograph.

E. Use of Cyclic AMP Phosphodiesterase to Destroy Cyclic AMP

Because authentic ^{3}H-cyclic AMP tracer was added to each unknown sample, there was the opportunity to verify the authenticity of the cyclic AMP peak in every sample. Each sample's specific activity depends upon the amount of tritiated tracer cyclic AMP per amount of endogenous cyclic AMP present (ready for injection into the chromatograph) was treated with phosphodiesterase to destroy about 50% of the cyclic AMP present in the sample. The treated sample was then injected into the chromatograph. If the cyclic AMP peak was authentic cyclic AMP, then the chromatographic peak height and radioactivity corresponding to the chromatographic peak should be reduced equally. That is, the specific activity of the sample should not change if the peak absorbancy was authentic cyclic AMP.

To accomplish this test in practice, a 10 μl fraction remaining from the purified tissue or urine sample was incubated for 10 min at 30°C with 10 μl of a solution containing 5 mM $MgCl_2$, 100 mM Tris-HCl, pH 8.0, and 1 μl of phosphodiesterase. The reaction was stopped in a boiling water bath for 2 min, and the treated sample then injected directly into the high-pressure liquid chromatograph.

F. Linearity and Reproducibility

The high-pressure anion exchange chromatographic assay for cyclic AMP is linear over a wide range of cyclic AMP concentrations. When standards of 0, 50, 100, 200, and 400 pmoles of cyclic AMP were carried through the complete assay procedure, 0, 49, 100, 195, and 396 pmoles of cyclic AMP were found, respectively.

The reproducibility of the method was checked by the quintuplicate analysis of a large sample of rabbit skeletal muscle which had been treated with epinephrine. The tissue (1,330 mg) was homogenized in 15 ml of 5% TCA containing 78,500 cpm of ^{3}H-cyclic AMP. The supernatant was then divided into five equal parts and each fraction carried through the assay procedure as described for tissue analysis. The purified samples, before injection into the chromatograph, were then taken up with 40 μl of HCl, pH 2.20, and 10 to 20 μl injected. The result of this experiment is shown in Table 1. The coefficient of variation is less than 5% for this method. The control level of cyclic AMP in rabbit skeletal muscle was 0.46 ± 0.06 μmoles/kg (S.E.M.).

G. Determination of ^{14}C-Cyclic AMP Specific Activity in Frog Ventricles Prelabeled with ^{14}C-Adenosine

The amount of cyclic AMP in a tissue extract is based upon the ultraviolet absorbance and amount of radioactive tracer nucleotide recovered

TABLE 1. *Reproducibility of cyclic AMP assay method checked in skeletal muscle*

Sample	Wt (mg)	μl Injected	Peak Height (mm)	cpm	Amount Injected (pmoles)	Total % Recovery	Cyclic AMP (pmoles)	μM cyclic AMP/kg
1	266	20	139	750	24.2	9.6	506	1.90
2	266	10	216	1,100	37.6	28.0	536	2.01
3	266	10	201	1,000	35.0	25.5	549	2.06
4	266	10	223	1,250	38.8	31.9	487	1.83
5	266	10	210	1,100	36.5	28.0	521	1.96
						mean		1.95
						standard deviation		0.09
						standard error of the mean		0.04
						coefficient of variation		4.6%

Reference = 15,700 cpm ^{3}H-cyclic AMP
40 pmoles cyclic AMP = 230 mm peak height

in the pure high-pressure anion exchange chromatographic peak. Since high specific activity tritiated cyclic AMP was used to determine purification recovery, experiments in which cyclic AMP becomes labeled *in vivo* by ^{14}C, ^{32}P, or ^{33}P precursors are easily performed. It has been shown in fat cells and in brain that labeled adenine or adenosine can be incorporated into ATP and also into cyclic AMP. Rall and Sattin (1970) first determined ^{14}C-cyclic AMP specific activity in brain slices. They were able to show that the specific activity of cyclic AMP was markedly increased when the prelabeled slices were treated with histamine. The determination of cyclic AMP specific activity is very difficult and requires using a combination of methods, and only Rall and Sattin (1970) and Krishna, Forn, Voigt, Paul, and Gessa (1970) have so far reported these measurements.

This new high-pressure chromatographic method was used to demonstrate that reliable ^{14}C-cyclic AMP specific activity could be easily determined in perfused frog ventricles after the adenine nucleotides were prelabeled with ^{14}C-adenosine. Frog ventricles were perfused, using Straub cannulae as previously described (Brooker and Thomas, 1971), and the cellular adenine nucleotides prelabeled for 90 min in 2 ml of bicarbonate buffer, containing 2 μC of ^{14}C-adenosine. Excess ^{14}C radioactivity was washed out from the ventricles by 15 changes of the bicarbonate buffer at 2 min intervals. The perfusion solution after the 15th wash had only background radiation. However, the ventricles had a mean of 14,000 cpm of ^{14}C per mg wet weight.

Cyclic AMP and ^{14}C-cyclic AMP was determined in ventricles prelabeled with ^{14}C-adenosine. The results of this experiment are summarized in Table 2. ^{14}C-cyclic AMP specific activity was about the same as ^{14}C-ATP specific activity, and appeared not to change during the first 30 min after the washout was completed. In addition, treatment with phosphodiesterase destroyed both ^{14}C and ^{3}H tracer cyclic AMP to the same extent, proving the authentic nature of the ^{14}C-cyclic AMP formed by prelabeling.

H. Measurement of Cyclic GMP

High-pressure anion exchange chromatography of guanosine 3',5'-cyclic monophosphate (cyclic GMP) requires a more concentrated buffer for elution from the high-pressure column. HCl, pH 2.25, containing 0.2 M NaCl works well for chromatography of cyclic GMP. Figure 8 shows a chromatogram of 50 pmoles of cyclic GMP. Prepurification of biological samples in a manner similar to that described for the measurement of cyclic AMP should make it possible to extend this high-pressure anion exchange chromatography method to the measurement of cyclic GMP in the same biological specimen in which cyclic AMP or other nucleotides are also being measured.

TABLE 2. *Determination of* ^{14}C-*cyclic AMP and* ^{14}C-*ATP specific activities in frog ventricles*

Exp. A.	No.	Mean $\pm$ S.E.M.
^{14}C-Cyclic AMP specific activity (cpm/pmole)	6	4.20 $\pm$ 0.41
^{14}C-ATP specific activity (cpm/pmole)	6	3.43 $\pm$ 0.14

Exp. B.		After Phosphodiesterase Treatment	
Heart No.	$\dfrac{^3\text{H-Cyclic AMP}}{^{14}\text{C-Cyclic AMP}}$	% Reduction in Tracer ^{3}H-Cyclic AMP	$\dfrac{^3\text{H-Cyclic AMP}}{^{14}\text{C-Cyclic AMP}}$
1	204	26	190
2	214	25	206
3	207	43	207
4	190	23	181
5	215	36	240
6	312	43	290

An aliquot of each prepurified sample was treated with phosphodiesterase and the ratio of tracer ^{3}H-cyclic AMP compared with the ^{14}C-cyclic AMP counts. The ratio remained constant, even though 23 to 45% of the tracer was destroyed by the phosphodiesterase. This proved the authentic nature of the ^{14}C-cyclic AMP formed in prelabeled frog ventricles.

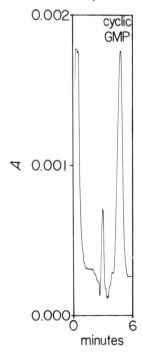

FIG. 8. High-pressure anion exchange chromatographic measurement of cyclic GMP. Cyclic GMP peak is 50 pmoles. Eluting buffer is HCl, pH 2.20, in 0.2 M NaCl. (From Brooker, 1971 *c*)

IV. DISCUSSION

It should be recognized that each cyclic AMP assay method has certain advantages and disadvantages. The method of choice of one laboratory may be unsuitable for another.

Any compound which alters the rate of the phosphodiesterase reaction could potentially interfere with the enzyme displacement assay. Because of this, it is important to verify that the tissue prepurification has eliminated any such substance. Confidence in a cyclic AMP level is increased if the unknown sample dilutes along the cyclic AMP standard curve and if cyclic AMP added to the sample is quantitatively recovered. Another validity check more commonly used is the destruction of all cyclic AMP with phosphodiesterase, denaturation of the phosphodiesterase, and reassay of the sample after the destruction of cyclic AMP. However, the zero cyclic AMP result after reassay is not rigorous proof that the original sample was authentic cyclic AMP. Because the phosphodiesterase contains other enzymatic activities, cyclic AMP and potentially interfering substances might also be eliminated from the sample after phosphodiesterase treatment. Thus the false impression that an assayed sample was cyclic AMP could be the result of such a validity check.

The high-pressure anion exchange chromatographic physical assay for cyclic AMP has proven to be reliable, consistently linear, and devoid of the many problems that have been associated with enzymatic assay methods of cyclic AMP. The presence of cyclic AMP in the chromatographic effluent is directly measured by the sensitive flow cell detector. Other methods probably incur more variation and error because the final measurement of cyclic AMP is less direct and is based upon the result of an enzymatic amplification, enzyme activation, or displacement of ^{3}H-cyclic AMP from a binding protein, enzyme, or antibody. Each of these methods has been successfully applied to the measurement of cyclic AMP in biological materials; however, each new type of biological specimen requires initial validity checks to insure the accuracy of the results obtained.

The high-pressure chromatographic and improved enzymatic isotope displacement methods have been used to measure cyclic AMP in a wide diversity of biological samples without any apparent interferences. The chances that the high-pressure chromatographic peak is a compound other than cyclic AMP seems remote because of the extensive prepurification. Bases or nucleosides are eliminated either by ether extractions, the first AG-1-X2 column, the Zn-Ba precipitation as originally described by Krishna, Weiss, and Brodie (1968), or by thin-layer chromatography. Nucleotides are eliminated by Zn-Ba precipitation, the AG-1-X2 column, and by thin-layer chromatography. In this thin-layer system, nucleotides remain at the origin and cyclic AMP has an R_f of about 0.4.

The biological methods for the assay of cyclic AMP suffer from a rather small measurement range and require dilution or concentration of the sample for optimal results. In contrast, the new high-pressure chromatographic assay has been found to be completely linear between 0 and 10,000 pmoles of nucleotide.

Another advantage of this new chromatographic system for cyclic AMP measurement is the unique ability to prove the authenticity, if desired, of each cyclic AMP measurement by the partial destruction with phosphodiesterase of an aliquot of the prepurified sample. Since the assay method is based upon a specific activity measurement, the enzyme must hydrolyze the nonradioactive and tracer cyclic AMP at the same rate for the specific activity to remain the same as that of the untreated sample. If a phosphodiesterase inhibitor were contributing to the absorbance of the cyclic AMP peak but was not degraded by the phosphodiesterase, then the reinjected sample would have a lower specific activity. If a phosphodiesterase inhibitor were contributing to the absorbance of the cyclic AMP peak and was also degraded by the phosphodiesterase, then the K_i would have to equal the K_m for cyclic AMP if the specific activity of ^{3}H-cyclic AMP were to remain constant. In addition, the inhibitor would need to co-chromatograph with cyclic AMP in all of the prepurification procedures. The presence of a competitive or noncompetitive phosphodiesterase inhibitor in the sample which does not contribute to the absorbance of the cyclic AMP peak would not interfere with the assay, of course, nor would it affect the cyclic AMP specific activity, since the hydrolysis of both the tracer cyclic AMP and the cyclic AMP from the tissue extract would be reduced equally.

The limits of detection of the high pressure chromatographic assay method are a function of the specific activity of the tracer used, the purification recovery, and the sensitivity of the ultraviolet flow cell detector, whereas the sensitivity limits of the enzyme displacement method are a function of the reaction volume, the K_m of the phosphodiesterase, and the isotope specific activity. ^{3}H-cyclic AMP with specific activities in the range of 25 C/mMole is now commercially available. Thus if 2,750 dpm of this tracer were used to determine recovery, then only 0.05 pmoles of cyclic AMP would be contributed to the sample by the tracer. This then makes it easy to measure cyclic AMP with good accuracy in amounts which are 5 to 10 times the amount of tracer added, e.g., 0.25 to 0.5 pmoles. Since purification recovery is about 25%, the tissue sample should contain 1 to 2 pmoles. Because the ultraviolet flow cell at present can measure 4 pmoles easily (10% of full scale peak) and purification recovery is 25%, a tissue or urine sample should contain about 15 pmoles for measurement by the high-pressure chromatographic method. This is easily done in most cases, since 50 mg of most tissues generally contains more than this amount of cyclic AMP. Thus the sensitivity limits of this method are now set by the sensitivity of the ultraviolet flow cell detector.

The sensitivity of the detector might be increased 5- to 10-fold by increasing the pathlength from 1 to 10 cm. Thus, tissue samples as small as 5 mg might be easily accommodated. Increased sensitivity is also possible by eluting cyclic AMP in a smaller volume with a higher concentration of HCl. Because the cyclic AMP peak would emerge earlier, additional precautions on prepurification of samples probably would be necessary to reduce the amount of ultraviolet absorbing substances in the breakthrough peak. If the breakthrough peak were not reduced, then a digital integrator with tangential baseline correction is useful to measure the cyclic AMP peak which would occur upon the sloping baseline from the breakthrough peak. These integration devices are widely used for this purpose in gas chromatographic applications and have also been useful in this application. Thus with an increased path length of the detector and a digital integrator, the sensitivity of the assay system should make high precision measurements possible in tissue samples containing 1 to 2 pmoles of cyclic AMP. Thus the detection limits would be equally limited by the tracer specific activity and the detector sensitivity. With the sensitivity already described in this chapter, most experimental samples are large enough for cyclic AMP analysis.

One factor which contributes to the high precision seen with the high-pressure chromatographic method is that the same sample is used to measure endogenous cyclic AMP and also counted to determine purification recovery. A small error in determining purification recovery can lead to a large deviation in the final result. For example, if the aliquot assayed for cyclic AMP were 25% of the original sample, and the sample counted for recovery deviated by only 1%, and was actually 24% of the sample (but assumed to be 25%), then the final result would be off by 4%; if the deviation were 3 or 4%, then the final result might deviate as much as 12 to 15%. With multiple pipeting of small volumes, it seems probable that this kind of deviation will occur. With the present high-pressure chromatographic assay, the assayed sample is the sample counted for recovery, since the peak is collected directly into a liquid scintillation vial, scintillant added, and the vial counted.

Another distinct advantage of this method is its usefulness for specific activity determinations. These determinations appear essential to determine the turnover rates, pool sizes, and effects of drugs, ions, and hormones upon cyclic AMP metabolism. Most other methods for cyclic AMP require prepurification of cyclic AMP before the final measurement. Use of ^{3}H-cyclic AMP tracer to determine recovery is difficult in the displacement assays, since high specific activity ^{3}H-cyclic AMP is generally used to compete with bound cyclic AMP in the assay. With the high-pressure ion exchange chromatography method just described, the assayed sample is simply counted for ^{3}H and ^{14}C to determine cyclic AMP recovery and ^{14}C-cyclic AMP specific activity.

ACKNOWLEDGMENTS

It is a pleasure to acknowledge the valuable help and advice of Harrison Frank, the expert technical assistance of Felicidad Avila and Sharon Laws, and the expert secretarial assistance of Patricia Lissner and Georgene Dennison. This research was supported by a Grant-in-Aid from the American Heart Association, The Los Angeles County Heart Association, the Diabetes Association of Southern California, U.S. Public Health Service Grant HE 13340, and a travel grant from Burroughs Wellcome, Inc.

REFERENCES

Breckenridge, B. McL.: Review of cyclic AMP assay methods. *Annals of the New York Academy of Sciences (in press).*

Brooker, G., Thomas, L. J., Jr., and Appleman, M. M. (1968): The assay of adenosine 3',5'-cyclic monophosphate in biological materials by enzymatic radioisotopic displacement. *Biochemistry,* 7:4177.

Brooker, G., and Appleman, M. M. (1968): The theoretical basis for the measurement of compounds by enzymatic radioisotopic displacement. *Biochemistry,* 7:4182.

Brooker, G. (1970): Determination of picomole amounts of enzymatically formed adenosine 3',5' cyclic monophosphate by high-pressure anion exchange chromatography. *Analytical Chemistry,* 42:1108.

Brooker, G. (1971*a*): Measurement of cyclic AMP in biological materials by high pressure anion exchange liquid chromatography with U.V. detection. *Federation Proceedings,* 30:140.

Brooker, G. (1971*b*): Effect of temperature control on the stability and sensitivity of a high pressure liquid chromatography ultraviolet flow cell detector. *Analytical Chemistry,* 43:1095.

Brooker, G. (1971*c*): High pressure anion exchange chromatographic measurement of cyclic 3',5' adenosine monophosphate and ^{14}C cyclic AMP specific activity in myocardium prelabeled with ^{14}C adenosine. *Journal of Biological Chemistry,* 246:7810.

Brooker, G., and Thomas, L. J., Jr. (1971): Phosphatase and ouabain-sensitive adenosine triphosphatase activities of the perfused frog heart. *Molecular Pharmacology,* 7:199.

Butcher, R. W., and Sutherland, E. W. (1962): Adenosine 3',5' phosphate in biological materials. I. Purification and properties of cyclic 3',5'-nucleotide phosphodiesterase and use of this enzyme to characterize adenosine 3',5'-phosphate in human urine. *Journal of Biological Chemistry,* 237:1244.

Horvath, C. G., Preiss, B. A., and Lipsky, S. R. (1967): Fast liquid chromatography: An investigation of operating parameters and the separation of nucleotides on pellicular ion exchanges. *Analytical Chemistry,* 39:1422.

Krishna, G., Weiss, B., and Brodie, B. B. (1968): A simple, sensitive method for the assay of adenyl cyclase. *Journal of Pharmacology and Experimental Therapeutics,* 163:379.

Krishna, G., Forn, J., Voigt, K., Paul, M., and Gessa, G. L. (1970): Dynamic aspects of neurohormonal control of cyclic 3',5'-AMP synthesis in brain. In: *Advances in Biochemical Psychopharmacology,* Vol. 3, edited by E. Costa and P. Greengard, p. 155. Raven Press, New York.

Newsholme, E. A., and Taylor, K. (1968): A new principle for the assay of metabolites involving the combined effects of isotope dilution and enzymatic catalysis. *Biochimica et Biophysica Acta,* 158:11.

Rall, T. W., and Sutherland, E. W. (1958): Formation of a cyclic adenine ribonucleotide by tissue particles. *Journal of Biological Chemistry,* 232:1065.

Rall, T. W., and Sattin, A. (1970): Factors influencing the accumulation of cyclic AMP in brain tissue. In: *Advances in Biochemical Psychopharmacology*, Vol. 3, edited by E. Costa and P. Greengard, p. 113. Raven Press, New York.

Robison, G. A., Butcher, R. W., and Sutherland, E. W. (1968): Cyclic AMP. *Annual Review of Biochemistry*, 37:149.

Smith, M., Drummond, G. I., and Khorana, H. G. (1961): Cyclic phosphates. IV. Ribonucleoside-3′,5′ cyclic phosphates. A general method of synthesis and some properties. *Journal of the American Chemical Society*, 83:698.

Sutherland, E. W., and Rall, T. W. (1960): The relation of adenosine 3′,5′-phosphate and phosphorylase to the actions of catecholamines and other hormones. *Pharmacological Reviews*, 12:265.

Thompson, J. W., and Appleman, M. M. (1971): Multiple cyclic nucleotide phosphodiesterase activities from rat brain. *Biochemistry*, 10:311.

Advances in Cyclic Nucleotide Research, Vol. 2
Raven Press, New York © 1972

Determination of Relative Levels of Cyclic AMP in Tissues or Cells Prelabeled with Radioactive Adenine

John W. Kebabian, Jyh-Fa Kuo, and Paul Greengard

Department of Pharmacology, Yale University School of Medicine, New Haven, Connecticut 06510

I. PRINCIPLE

Various mammalian tissues and cells, when incubated in a solution containing adenine, rapidly concentrate this purine and convert it into 5'-AMP, and subsequently into ATP. Several reviews have summarized the recent literature on these reactions (for example, Murray, Elliott, and Atkinson, 1970; Murray, 1971). Kuo and his colleagues (Kuo and Dill, 1968; Kuo, 1969; Kuo and DeRenzo, 1969), using isolated adipocytes, were the first to report that this newly synthesized ATP is actively converted into adenosine 3',5'-monophosphate (cyclic AMP); many investigators, using a variety of tissues and cells, have subsequently made similar observations (e.g., Shimizu, Creveling, and Daly, 1969; Krishna, Forn, Voigt, Paul, and Gessa, 1970; Kuehl, Humes, Turnoff, Cirillo, and Ham, 1970; Chasin, Revkin, Mamrak, Samiego, and Hess, 1971). The basis of the prelabeling technique is that, in tissues and cells pulse-labeled with radioactive adenine, the relative amounts of cyclic AMP newly formed from ATP can be determined by isolating the cyclic nucleotide and measuring the amount of radioisotope which it contains. It is not possible to calculate the absolute amount of the cyclic nucleotide in a sample by this method. However, in a number of different experimental situations, it has been found that changes in the level of radioactive cyclic AMP reflected changes in the absolute levels of the cyclic nucleotide. The prelabeling technique provides a convenient method for studying the dynamic changes in cyclic AMP levels in tissues and cells.

The prelabeling technique gives an accurate and sensitive estimation of the relative levels of radioactive cyclic AMP in cellular systems. The limitation on the measurement of small amounts of cyclic AMP with an acceptable precision, i.e., the limit of sensitivity of the assay, is a function of the specific activity of the radioactive adenine utilized for the prelabeling procedure. Thus, by increasing the specific activity of the radioactive adenine, proportionately smaller amounts of cyclic AMP can be measured with the same precision.

II. PROCEDURE

A. Prelabeling the Tissues or Cells

In experiments in which changes in tissue cyclic AMP levels are to be studied, tissues should be prepared so that the uptake of adenine is not seriously limited by the rate of diffusion of the purine into the interior of the tissues. For example, neural or cardiac tissues may be prepared as slices or small pieces; in contrast, distinct cells, such as adipocytes, or small groups of cells, such as pancreatic islets, can be isolated and used as such. The selection of the concentration of adenine utilized for prelabeling will be influenced by the anticipated dilution of the newly formed nucleoside and nucleotides by endogenous compounds. Furthermore, the duration of the period of prelabeling may vary for different tissues, depending on the efficiency of the uptake of the adenine and the rate of ATP synthesis. In studies with slices of either neural (Shimizu et al., 1969; Kebabian and Greengard, 1971) or cardiac tissues (Lee, Kuo, and Greengard, 1971), a 45-min incubation of the tissue slices in Krebs-Ringer bicarbonate buffer[1] fortified by the addition of 10 mM glucose (KG), at 37°C, containing 10 to 40 μM [8-^{14}C] adenine (41.6 mC per mmole) or an equivalent concentration of [2-^{3}H] adenine, resulted in adequate labeling of the ATP pool of the tissue. Conditions similar to those utilized in these experiments have also been shown to be satisfactory for prelabeling both rat adipocytes (for example, see Kuo, 1969) and rat pancreatic islets (Kuo, Kuo, Hodgins, and Greengard, 1972). The presence of glucose in the incubation medium facilitates the incorporation of radioactive adenine into the cellular ATP pool. In typical experiments, slices of the ventricle of rat heart incorporated about 15% of the adenine which had been added to the medium, and about 0.1% of this was found as cyclic AMP. Likewise, rat adipocytes incorporated about 17% of the extracellular adenine; however, about 12% of this adenine was converted to cyclic AMP.

At the end of the period of prelabeling, the adenine in the extracellular fluids must be removed. This is easily accomplished if the tissue slices are incubated in a special vessel made of a large bore (5-cm diameter) plastic

---

[1] Krebs-Ringer bicarbonate buffer (pH 7.4) contained (in mmoles per liter) NaCl, 122; KCl, 3; CaCl$_2$, 1.3; MgSO$_4$, 1.2; KH$_2$PO$_4$, 0.4; NaHCO$_3$, 25. A mixture of 95% O$_2$ and 5% CO$_2$ was bubbled through the solution for at least 1 hr prior to the experiment.

tube; one end is covered with a nylon mesh, and a wire bail is attached to the other end. The vessel fits into a larger beaker which contains the solution of adenine in KG. At the end of the incubation, the special vessel is removed and the tissue is washed by pouring KG over the pieces of tissue, which are retained by the nylon mesh. The vessel containing the washed tissue is then put into another beaker of KG until aliquots of the tissue are required for incubation. This procedure minimizes any physical damage to the prelabeled tissue, while permitting a rapid and efficient removal of the medium containing the radioactive adenine. When isolated cells are prelabeled, these cells can be separated from the incubation medium by a brief centrifugation of the incubation mixture followed by aspiration. Extracellular radioactive adenine remaining is removed by washing the cells in an appropriate volume of fresh buffer, which is removed by a second centrifugation.

B. Incubation of Tissue Slices or Isolated Cells

The conditions employed to study the effects of various agents on the levels of cyclic AMP may vary either because of the tissue or cell types studied, or because of the experimental design. In studies with the bovine superior cervical ganglion (Kebabian and Greengard, 1971), the incubations were performed in 12-ml homogenizers containing 5 ml of KG with 10 mM theophylline plus test substances. The tissue was kept from settling to the bottom of the homogenizer by bubbling a mixture of 95% O_2 and 5% CO_2 through the solution. The incubation was terminated by allowing the tissue to settle to the bottom of the homogenizer, rapidly removing the medium, and adding 1.5 ml of 6% trichloracetic acid (TCA), 4°C, to the tissue.

C. Purification of the Cyclic AMP

The tissue is homogenized, and 0.1 μmole (approximately 3,000 cpm) of either cyclic AMP-^{3}H or cyclic AMP-^{14}C, depending on the isotope utilized for prelabeling, is added to the homogenate. This cyclic AMP both serves as carrier and allows the calculation of the recovery of the cyclic AMP for each sample. When isolated cells are used, they are ruptured conveniently with a sonicator. The homogenate is centrifuged and the supernatant is separated from the pellet. The amount of protein in the TCA-insoluble material is determined by the method of Lowry, Rosebrough, Farr, and Randall (1951).

The labeled cyclic AMP is separated from the other metabolites of the labeled adenine by using the procedure of Krishna, Weiss, and Brodie (1968) with a few modifications. The TCA either may be removed from the supernatant by extraction with diethyl ether, or may be neutralized by the addition to the supernatant of an appropriate volume of 1 M tris-hydroxymethylaminomethane. Initially, radioactive nucleotides other than cyclic AMP are removed from the solution by the $ZnSO_4$–$Ba(OH)_2$ procedure of Krishna, Weiss, and

Brodie (1968) which has been extensively investigated by Shimizu, Creveling, and Daly (1969). We have modified their procedure slightly. Equal volumes of a 0.3 N solution of $ZnSO_4$ and a saturated solution of $Ba(OH)_2$ are mixed together; white, insoluble $BaSO_4$ forms immediately. Three 0.2-ml aliquots of this slurry are added to a 1.0-ml aliquot of the TCA-free extract; after each addition the mixture is vigorously agitated with a vortex mixer. After the three additions, the mixture is centrifuged. We have found that the use of the $BaSO_4$ slurry eliminated the problems caused by the conversion of ATP to cyclic AMP observed when the two reagents were added separately. A second purification step for the cyclic AMP is required if, after the $BaSO_4$ treatment, an appreciable amount of contaminating material remains in the supernatant solution. In this case, aliquots (usually 0.5 ml) of the supernatant from the $BaSO_4$ step are charged onto a column (0.5 × 5.0 cm) of Dowex AG50W X8 (H^+ form, 100–200 mesh, BioRad) which has previously been equilibrated with 1 mM phosphate buffer, pH 7.0 (5 to 10 ml). These columns are made from disposable Pasteur pipettes by placing a number 4 cotton pellet in the bottom; each column is filled by delivering an aqueous suspension of the resin from a large bore syringe with a long stainless steel needle. The charged column is then washed with 4.0 ml of 1 mM phosphate buffer, pH 7.0, and the cyclic AMP eluted with 5.0 ml of H_2O. It is imperative that the elution pattern of cyclic AMP be determined for each lot of resin because variations do exist between different lots of the resin. Either aliquots of the eluate containing the cyclic AMP are counted directly or the entire eluate may be evaporated to dryness, redissolved in an aliquot of water, and the radioactivity counted after addition of 10 ml of scintillation fluid (prepared by dissolving 4 g of Omnifluor, New England Nuclear, in 1 liter each of toluene and ethylene glycol monoethyl ether).

The simultaneous determination of the amount of ^3H and ^{14}C in each sample permits the computation of the total amount of cyclic AMP-^{14}C which the original tissue sample contained. The total amount of radioactive cyclic AMP in each sample is normalized for the variations in the amount of tissue by expressing the data as total counts per minute of cyclic AMP per mg protein. The recent article by Kobayashi and Maudsley (1970) has dealt with both the practical and the theoretical aspects of double label counting.

For any given tissue or cell preparation, it is desirable to compare the amount of radioisotope which the cyclic AMP contains with the absolute levels of the cyclic nucleotide (e.g., using one of the methods detailed elsewhere in this volume). The results of such a comparison for either ganglionic tissue (Table 1) or cardiac tissue (Table 2) demonstrate that the radioactive cyclic AMP formed reflects the changes of the absolute levels of the cyclic nucleotide. Thus, the specific activity of the newly formed cyclic AMP (and presumably of the ATP pool which serves as the precursor for the cyclic AMP) was the same in the presence as in the absence of the stimulatory hormones.

TABLE 1. *Levels of cyclic AMP in the bovine superior cervical sympathetic ganglion*

Incubation conditions	Cyclic AMP found by	
	Prelabeling method (cpm/mg protein)	Protein kinase method (pmoles/mg protein)
Unincubated	85.4 ± 5.5	14.5 ± 0.55
Control	210.5 ± 11.2 (3.2)	41.0 ± 0.81 (2.6)
Dopamine, 30 μM	672.0 ± 19.7 (9.1)	119.0 ± 3.4 (9.7)

One-half ml of the trichloracetic acid extract (total 1 ml) of the tissue from the prelabeling experiment was used for the measurement of cyclic AMP level by the protein kinase catalytic method (Kuo and Greengard, 1970). Each value represents mean ± S.E.M. of quadruplicate incubations. The numbers in parentheses indicate fold increases in cyclic AMP levels relative to unincubated tissue. (From Kebabian and Greengard, 1971.)

TABLE 2. *Increase in rat ventricular cyclic AMP stimulated by isoproterenol and glucagon*

Incubation conditions	Cyclic AMP found by	
	Prelabeling method (cpm/mg protein)	Protein kinase method (pmoles/mg protein)
Control	363 ± 16	1.42 ± 0.42
Isoproterenol, 10 μM	2691 ± 189 (7.4)	8.18 ± 0.56 (5.8)
Glucagon, 10 μM	2389 ± 72 (6.6)	8.03 ± 1.46 (5.7)

Two-tenths ml of the trichloracetic acid extract (total 1 ml) of the tissue from the prelabeling experiment was used for the measurement of cyclic AMP level by the protein kinase catalytic method (Kuo and Greengard, 1970). Each value represents mean ± S.E.M. of triplicate incubations. The numbers in parentheses indicate fold increases in cyclic AMP levels due to hormones. (From Lee, Kuo, and Greengard, 1971.)

III. CONCLUDING REMARKS

The prelabeling technique is a simple and accurate method for the determination of the relative levels of cyclic AMP in tissue and cells. Certain other methods can measure the absolute levels of cyclic AMP in tissues and cells, and also in cell-free systems, with comparable simplicity and accuracy. However, when very limited amounts of tissues are available, the prelabeling technique, with its great sensitivity, can be uniquely useful for studying dynamic changes in the levels of cyclic AMP.

By analogy with the method described above, which is based on prelabeling the ATP pool with radioactive adenine, it should be possible to prelabel the GTP pool with radioactive guanine and study the accumulation of radioactive cyclic GMP derived from this nucleotide. Since cyclic GMP is present in mammalian tissues, and some agents are known to specifically alter cyclic

GMP levels, it seems desirable that a simple, yet sensitive, prelabeling technique be developed for the cyclic GMP system. Attempts to develop such a method are in progress in our laboratory.

ACKNOWLEDGMENTS

This work was supported by U.S. Public Health Service Grants NS-08440, MH-17387, and HE-13305, and by Grant GB-27510 from the National Science Foundation. One of us (JFK) is the recipient of U.S. Public Health Service Research Career Development Award 1 K4 GM-50165.

REFERENCES

Chasin, M., Rivkin, I., Mamrak, F., Samaniego, S. G., and Hess, S. M. (1971): α- and β-Adrenergic receptors as mediators of accumulation of cyclic adenosine 3′,5′-monophosphate in specific areas of guinea pig brain. *Journal of Biological Chemistry*, 246:3037–3041.

Kebabian, J. W., and Greengard, P. (1971): Dopamine-sensitive adenyl cyclase: Possible role in synaptic transmission. *Science*, 174:1346–1349.

Kobayashi, Y., and Maudsley, D. V. (1971): Practical aspects of double isotope counting. In: *The Current Status of Liquid Scintillation Counting*, edited by E. D. Bransome, Jr. Grune and Stratton, New York.

Krishna, G., Forn, J., Voigt, K., Paul, M., and Gessa, G. L. (1970): Dynamic aspects of neurohormonal control of cyclic 3′,5′-AMP synthesis in brain. In: *Role of Cyclic AMP in Cell Function*, edited by P. Greengard and E. Costa. Raven Press, New York.

Krishna, G., Weiss, B., and Brodie, B. B. (1968): A simple, sensitive method for the assay of adenyl cyclase. *Journal of Pharmacology and Experimental Therapeutics*, 163:379–385.

Kuehl, F. A., Humes, J. L., Turnoff, J., Cirillo, V. J., and Ham, E. A. (1970): Prostaglandin receptor site: Evidence for an essential role in the action leuteinizing hormone. *Science*, 169:883–886.

Kuo, J. F., and Dill, I. K. (1968): Antilipolytic action of valinomycin and nonactin in isolated adipose cells through inhibition of adenyl cyclase. *Biochemical and Biophysical Research Communications*, 32:333–337.

Kuo, J. F. (1969): Effects of deoxyfrenolicin on isolated adipose cells. II. Lipolysis, adenosine 3′,5′-monophosphate levels and comparison with the effects of vitamin K_5. *Biochemical Pharmacology*, 18:757–766.

Kuo, J. F., and DeRenzo, E. C. (1969): A comparison of the effects of lipolytic and antilipolytic agents on adenosine 3′,5′-monophosphate levels in adipose cells as determined by prior labeling with adenine-8-^{14}C. *Journal of Biological Chemistry*, 244:2252–2260.

Kuo, J. F., and Greengard, P. (1970): Cyclic nucleotide-dependent protein kinases. VIII. An assay method for the measurement of adenosine 3′,5′-monophosphate in various tissues and a study of agents influencing its level in adipose cells. *Journal of Biological Chemistry*, 245:4067–4073.

Kuo, W. N., Kuo, J. F., Hodgins, D. S., and Greengard, P. (1972): Increase of adenosine 3′,5′-monophosphate in intact pancreatic islets by various insulin-releasing agents. *In preparation*.

Lee, T. P., Kuo, J. F., and Greengard, P. (1971): Regulation of myocardial cyclic AMP by isoproterenol, glucagon and acetylcholine. *Biochemical and Biophysical Research Communications*, 45:991–997.

Lowry, O. H., Rosebrough, N. J., Farr, A. L., and Randall, R. J. (1951): Protein measurement with the folin phenol reagent. *Journal of Biological Chemistry*, 193:265–275.

Murray, A. W., Elliott, D. C. and Atkinson, M. R. (1970): Nucleotide biosynthesis from preformed purines in mammalian cells: Regulatory mechanisms and biological significance. *Progress in Nucleic Acid Research and Molecular Biology*, 10:87–119.

Murray, A. W. (1971): The biological significance of purine salvage. *Annual Review of Biochemistry*, 40:811–826.

Shimizu, H., Creveling, C. R. and Daly, J. (1969): A radioisotopic method for measuring the formation of adenosine 3′,5′-cyclic monophosphate in incubated slices of brain. *Journal of Neurochemistry*, 16:1609–1619.

Name Index

Subject Index

141